Renewable Energy for Your Home

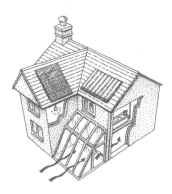

Renewable Energy for Your Home

Using Off-Grid Energy to Reduce Your Footprint, Lower Your Bills and Be More Self-Sufficient

**ALAN AND GILL
BRIDGEWATER**

Ulysses Press

Published in the U.S. by: Ulysses Press
P.O. Box 3440
Berkeley, CA 94703
www.ulyssespress.com

First published in 2008 in the U.K. as *The Off-Grid Energy Handbook* by New Holland Publishers Ltd.

Printed in Canada by Webcom

10 9 8 7 6 5 4 3 2 1

ISBN 978-1-56975-568-6
Library of Congress Control Number 2008911761

Acquisitions Editor: Nick Denton-Brown
Managing Editor: Claire Chun
U.S. Editor: Lauren Harrison
Production: Judith Metzener
Cover design: what!design @ whatweb.com
Interior design: AG&G Books

PHOTOGRAPHS
Center for Alternative Technology (CAT) (page 25)
Iskra Wind Turbines Ltd (page 2)
Solar Research Design Sdn. Bhd. (and page 9)

IMPORTANT NOTE TO THE READER
Please be sure to read the following important note before using any of the information in this book!

This book is a general introduction to the subject of renewable energy systems, and the reader is strongly cautioned to take note of the following warnings before attempting to use or relying on any of the information in this book.

The installation and operation of renewable energy systems involves a degree of risk and requires a degree of skill. It is the reader's responsibility to ensure that he or she understands all such risks and possesses all necessary skills. It is also the reader's responsibility to ensure that all proper installation, operation, regulatory, and safety rules are followed. It is also the reader's responsibility to obtain any and all necessary permits, and to follow all local laws and building regulations. Waste products resulting from the energy process are dangerous and may be subject to regulations under environmental or waste management programs in your area. Ensure that all environmental waste processing regulations are followed.

The reader may need additional information, advice, and instruction not included in this book before safely and lawfully installing and operating a renewable energy system.

Any decision to use the information in this book must be made by the reader on his or her own good judgment. This book is sold without warranties or guarantees of any kind, and the author and publisher disclaim any responsibility or liability for personal injury, property damage, or any other loss or damage, however caused, relating to the information in this book.

CONTENTS

INTRODUCTION

When Gill and I met at art school in the early 1960s, everything was fresh and anything was possible. We were just kids really, only 18, but we had a good idea of where we were going. Drawing inspiration from various hippy groups and communes, and generally being wound up by the talk about how oil supplies were running out, we decided that we wanted to go back to self-sufficient, off-grid energy basics. We were absolutely convinced that the big hand on the "dooms day energy clock" was just about to strike midnight. To cut a long story short, we put an ad in the newspaper that read "Isolated cottage wanted—must be set in its own grounds." One moment we were living in town in a little cottage complete with frilly curtains, a refrigerator, a gas stove, and just about every surface nicely decked out with flowery formica, and the next moment we were living in a red-brick ruin of a house in the middle of a field, with no water, electricity, sewage, or money—in fact nothing apart from a great deal of space, silence, darkness, and a huge feeling of peace. It was wonderful. I remember one night in early summer lying flat on my back in the meadow, not a house light or car light to be seen, and looking up at the stars—beautiful!

It may sound like something of a nightmare—unbelievably romantic and naïve, certainly—but at the time we intended to rebuild the house and incorporate various off-grid energy systems so we could be as self-sufficient as possible. Our master plan went something like this: find the old well and sort out the water supplies, put up a wind turbine, build a methane plant following the now-famous Harold Bates chicken-powered designs, put up various solar rooms and Trombe walls as described in *The Autonomous House* by Robert and Brenda Vale, get animals, and so on, until we had established our own little self-contained, self-sufficient, land-based, "space-pod" paradise.

So, here we are half a lifetime later, and where are we at? Well, the planet's oil supplies are on their last dribble, and people are just beginning to wake up to the fact that we are on the edge of a do-or-die catastrophe—do something about the energy crisis or face doom. Back in the 1970s, any mention of an energy crisis that would bring the world to its knees was written off as weird talk, and our little back-to-basics strivings were maybe something of a laugh, but not anymore. Now, according to the vast majority of the world's scientists, the human race is sitting and twiddling its thumbs on top of a time bomb that is just about to reach its last tick. The fossil fuels are running out, there are more cars than ever, our night sky is lit up like a Christmas tree, the forests are being hacked down faster than ever, the planet is overheating, governments are beginning to argue about energy, energy prices are doubling every few years…tick, tick, tick, tick.

That is enough of the "doom and gloom," but we do all have to get off our backsides and start, at the very least, to think about how we are going to use our energy. What are you going to say to your children and grandchildren when they point their fingers and ask what you did in the energy war? Some people have changed over to using low-energy light bulbs, but that just won't be enough. The exciting thing is that, whereas in the 1960s and '70s it was not really feasible to go down the off-grid route, other than to put up a wind turbine (and a pretty basic hit-or-miss machine at that), now everything is possible. There are wind turbines in just about every shape and size—solar-powered photovoltaic cells that will light up a whole home, geothermal systems that will both heat and cool your home, solar-powered water heaters, systems that turn manure into biogas and used corn oil into car fuel, and much more.

The good news is that we can now all do our part to make it better. I am not saying that you should rush out and physically sever your connections with the utility companies, but you could perhaps work toward putting up a solar-powered collector to heat your water or maybe an array of photovoltaic cells to cut your electricity bill; there are lots of very exciting and dynamic off-grid options out there. It is known that Queen Elizabeth II, Elton John, Daryl Hannah, and a whole host of other celebrities are going down the off-grid-energy road by installing their own geothermal and solar systems. Do they know something you don't? Well not now, because this book will show you how to install your very own off-grid energy systems. Just think about it—lowering energy bills, taking control, doing your part for the planet, not being the only person sitting in the dark when the lights go out.

So, if ever there was a time to roll up your sleeves and take control of your own fate, this is it. Are you going to be a leader or a follower? Are you going to passively accept the impending crisis, or are you going to do something about it? Just imagine how exciting it would be to have your very own renewable, off-grid energy systems! We hope that you find this book both helpful and inspirational.

Getting started

WHAT IS OFF-GRID ENERGY?

The dictionary defines "off-grid" as not being connected to power lines—water, electricity, gas, oil—and not using energy that is derived or produced from systems that are provided by companies. So, if you are living in a house that is remote from utility services, that derives its energy from the wind, the rain, or the sun, with no oil, bottled gas, bottled water, or batteries, then you are living off-grid. In many ways, off-grid living and self-sufficiency are the same. For example, if you collect your own rainwater, you could be described as being **self-sufficient in water**; if you have your own wind turbine, you are **self-sufficient in electricity**.

WHY LIVE OFF-GRID?

Most people set out on the off-grid route for one of four reasons:

1. They are tired of paying the utility companies and want to cut costs.

2. They have some sort of ethical, social, philosophical, or religious anxiety that is pushing them toward off-grid.

3. They see off-grid as being some sort of romantic challenge.

4. They live in such a remote area that they have no other choice.

Many off-grid beginners are motivated by what has come to be called "pioneer survivalism," meaning that they are moved by worries about some sort of doom scenario such as food running out, pollution or poison in our food and water, or a climate catastrophe. Others opt for off-grid living because they think the "hermit" way is a good method of taking back control. Off-gridders are sometimes referred to as "plug-pullers."

COUNTING THE COST

One of the difficulties with going off-grid is the initial cost. If you are living in a completely plugged-in house, you must either start fresh in a new home, or gradually "pull the plug" in your current home as items break down. In the long run, however, off-grid systems do cost less to run.

IS IT POSSIBLE TO BE TOTALLY OFF-GRID?

Most of our off-grid options—wind turbines, hydro turbines, solar cells—have their origins in on-grid industries, so we must accept the fact that we start out on the off-grid adventure with some amount of carbon, pollution, social, and environmental debts, and then do our best to keep the debt to a minimum. When Robert and Brenda Vale said in their book *The Autonomous House* that there is no such thing as a completely self-sufficient home, they were right on. Much the same goes for an off-grid home. We can work toward it, we can reduce our grid link-up, and we can reduce the impact of our carbon start-up debt, but we can never achieve 100 percent off-grid.

Farmhouse complete with wind turbine and solar heater

WORKING OUT YOUR REQUIREMENTS

You can calculate your "on-grid" electricity usage by totalling your household electric meter readings over a 12-month period, adding up the total kilowatt-hours (kWh) you have used, dividing the total by 365 to give you the average per day, and then dividing the result by 24 to give you a figure per hour. This tells you how much electricity you would need to generate should you go off-grid and puts the generating capacities of wind turbines and photovoltaic cells into perspective. It is unlikely that you can generate all the electricity that you need, especially if you use an electric stove, washing machine, dryer, or dishwasher, which use far more power than can realistically be generated or stored in batteries at home using modern technology. For the moment the answer is to generate as much supplementary electricity as you can, try to use less electricity, and consider alternative systems.

WHAT WILL IT POWER?

Most people interested in off-grid options take one look at a wind turbine or a solar heater, and immediately ask what it will power, how much it will power, and if it will do everything. There is no single off-grid system that will do everything in a modern home, so you will have to go for a mix of systems. You could have wood fires for heating, solar heaters for hot water, and a wind turbine for electricity; or PV cells for lighting, geothermal for heating, and biodiesel for cooking. It needs careful consideration. An off-grid system will always be a bit of a mix-and-match. You will be forced into thinking fresh, and you will have to spend time, effort, and money getting it right. Are you up to the challenge?

OFF-GRID OPTIONS

- **WIND TURBINES**

  A wind turbine will give you electrical energy to power one thing, and then, depending upon the size of your system, anything else that you want to power with electricity. For example, you could use the turbine for lighting and for running electrical appliances such as TV, radio, fan, computer, washing machine, and refrigerator. As for whether you use the DC 12-volt power directly to run 12-volt appliances, or turn the DC to AC household power, you must choose what is best for your own scenario.

- **HYDRO TURBINE**

  A hydro-turbine system is similar to a wind turbine in that it will give you electricity in the to power one thing, and will supply power for lights, small appliances, and a refrigerator—but it does not generate enough power to run a kitchen stove.

- **PASSIVE SOLAR HEATING**

  If you live in a moderate to warm climate, and you get it right, a passive solar system will enable you both to heat and to cool your rooms.

OFF-GRID OPTIONS (CONTINUED)

- ### SOLAR HEATERS
  Solar heaters will heat both space and water. There are all kinds of designs, some very good and some terrible. They all work on much the same principle: The sun shines on a heat-absorbent surface, water within the system heats up and starts to move, the hot water is stored in a tank, the water within the tank is used or cools down and begins to circulate, and so on around the system.

- ### PHOTOVOLTAIC CELLS
  Photovoltaic cells, sometimes called PVs, are in my mind the big promise for the future. The sun shines on the cells, which produce a minute amount of electricity; this energy is either used directly for little fans or motors or stored in a bank of batteries. PV cells can, depending upon the size and type, give you as much electrical energy as you need.

- ### GEOTHERMAL SYSTEMS
  A geothermal system will provide both heating and cooling.

- ### BIOGAS
  If you have the right setup—lots of space and plenty of suitable animal manure— this is an option that could power everything. You could use the gas to run a generator and the generator to run everything else.

- ### BIODIESEL SYSTEMS
  A biodiesel system could be used to run all manner of engines, for example a car or a generator.

- ### WOOD-FIRED SYSTEMS
  A wood-burning boiler could give you both heating and hot water, and also run your kitchen stove.

LOCATION OPTIONS

When it comes to setting up an off-grid energy system, location is everything. The right location makes everything possible, but in the wrong location you will be held up by regulations. Public perceptions about green issues are getting better, however, and local codes are becoming more accepting. If you live in an area where there are lots of buildings and a large population, your off-grid plans may be thwarted; if you live in a country area with lots of open space, you will probably be able to fulfill your off-grid dreams. This does not mean that you cannot go off-grid in densely populated areas, it just means it will be a little more difficult.

DEALING WITH OBJECTIONS

The problem with many off-grid structures and systems is that they are strange and different. While some people see large wind turbines as dynamic, elegant, and futuristic, others see them as ugly and out of proportion. This is where location comes in. If you live on a property in the middle of the woods, and your structure conforms to the various laws and codes, the installation will be that much easier because there is nobody to object.

If you live in a small town that has a neighborhood-watch committee, there is a good chance that someone will make objections, and if you want to put up a wind turbine on an urban street your neighbors will probably be suspicious. Be open, follow the rules, and try to encourage discussion. Don't be pushy or try to convert people to your way of thinking—just proceed quietly with your plans.

Wind turbine rotor options

FREQUENTLY ASKED QUESTIONS

- **Will a wind turbine be too noisy for my small-town house?** A modest-sized domestic turbine is no noisier than a small car. Check the noise levels when picking your turbine.

- **Will a hydro turbine pollute the water?** Normally there should be no pollution apart from the rare accidental drop of oil. A badly placed turbine, however, could damage the fish and perhaps churn up the mud or erode the river or stream banks. Take advice from a local specialist.

- **Will a geothermal system be too noisy for our townhouse?** The initial installation (drilling the borehole) will be noisy and the heat exchange will make much the same noise as a freezer, but that is all. A geothermal system will be silent when running.

- **Will a bank of batteries be a fire hazard?** Batteries are, by their nature, something of a risk because they contain acid, give off toxic fumes, and are flammable. Keep them in a well-insulated and ventilated shed, post warning signs, conform to health and safety codes, and install quality locks on the doors.

- **Will a wind turbine be dangerous?** There is the risk that bits could fall off and/or the tower could fall down. With such potential problems in mind, make sure that it will not fall on anyone's house, car, or garage, the road or pavement, or anything else. Depending upon the size of the turbine and your location, remember that you might need city or county permits.

- **Will a biogas system be too smelly?** It will almost certainly be smelly—you will be pouring buckets of manure in and shoveling buckets of slurry out. Make sure it is far away from your neighbors and generally downwind from the community. Try to build a system that involves and serves everyone in the local area. They all contribute manure, they all get the gas, and they all get whatever smells there are.

SCENARIOS

Urban scenario

In the context of off-grid energy, the urban scenario—meaning cities, towns, and densely-populated areas—is arguably the most challenging. It is not that off-grid options do not work in towns and cities—there are lots of city dwellers who have boreholes for geothermal systems, solar heaters on the roof, and solar photovoltaic panels on the walls—but all the necessary rules and regulations that are needed to make this large population of people run smoothly are that much more draconian. These codes are not exactly written in blood, but they are strict and they have to be followed. Ignorance is no excuse under the law, and if you break the rules it will not be easy.

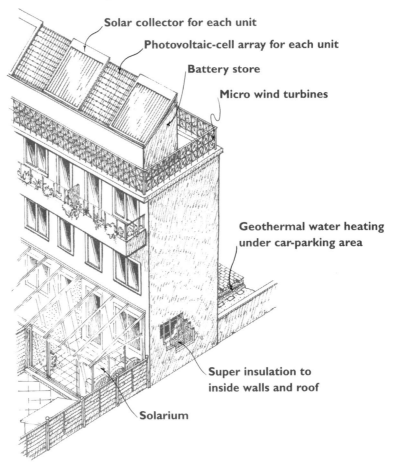

Solar collector for each unit

Photovoltaic-cell array for each unit

Battery store

Micro wind turbines

Geothermal water heating under car-parking area

Super insulation to inside walls and roof

Solarium

Urban roof detail

CODES AND LAWS

If you are living in a city and planning, for example, to put a solar panel on your roof, the first thing to do is check with the local authorities that you are not breaking any codes. In case you think you can get around or bend these codes—you can't. You may try openly to change them (see below), but you can't warp them.

RULES ARE MADE TO BE BROKEN

If you really do want to go ahead with a certain off-grid project and see that there is a code that says you can't, then the best way forward is to see if you can persuade the authorities to change their rules. For example, when one man decided that he wanted to keep chickens in his city garden, and the ever-vigilant neighborhood committee said that he could not, he joined forces with one neighbor, then another, and then another, and now the community has an inner-city farm.

NEIGHBORS

You either love them or you hate them. My advice, if you live in a city and don't get along with your neighbors, is to move to a cabin in the country. If you do live in a populated area and you want to stick with it, however, then make friends with your neighbors and be considerate. For example, if you want to build a rooftop balcony complete with an array of solar panels, what happens if some part of your project obstructs or lessens your neighbors' view? Such problems need to be addressed at the outset to avoid conflict.

NOISE

Noise can be a problem. In the same way that you don't want to be listening to your neighbors' car revving up, or their baby crying, or their heating system clunking away, your neighbor doesn't want to hear your turbine swishing around. Keep noise levels to an absolute minimum.

HEALTH AND SAFETY

You must make sure that every aspect associated with your system is noninjurious to health and completely safe. Fire, gases, noise, movement, and smells must all be carefully considered.

COMMUNITY PROJECTS

If you do not have room in your yard for a geothermal system or a wind turbine, for example, how about if you and your neighbors join forces? Everything is possible. Community projects range from geothermal systems under parking lots and solar collectors on rooftops to boreholes in front yards and micro wind turbines on balconies. If you want to change things on your block or street, you could consider a community project.

Small-town scenario

The small-town scenario is good and bad in equal parts. The main problem is that, while the space gives you the chance to go ahead with your off-grid plans, most of the eyes within the community, because it is small and insular, will be following your every move. It is not as difficult if you live on the edge of a small town as it is if you live right in the center. Just as in the city, you will have to get along and do your best to ensure that your plans do not upset people.

THE TYPE OF COMMUNITY

In my experience, the size, shape, and character of a small-town community are everything. Living on the edge of the town is very different from living next to the church, and living in an inland village is completely different from living by the sea. This being so, you must think very carefully before you go ahead with a project that might be considered contentious.

LOCAL CUSTOMS AND TRADITIONS

Your plans might annoy a neighbor—you can live with that—but they must not be built in the face of local customs and traditions. So, for example, if the river that runs through your property and along through the town is valued as a wildlife habitat for fishing or as a venue for water sports, it might not be wise for you to set up a hydro turbine. You might be following local codes, but are you going to be happy if you are shunned by half the town because some part of your plan does not fit?

NEIGHBORS

If you live in the city, there are so many people that you can ignore one or two without causing offense, but this is not so if you live in a small town where moods, feuds, and relationships are exaggerated by close contact. A good option, if you can manage it, is to involve your neighbors in your plans. For example, there is a little row of condos in a rural town where they have formed a community group. So far, they have built a geothermal system that runs through all the yards, and put solar panels up on a piece of jointly-owned land. They are all happy, and they are strong in their commitment to their joint ventures to the extent that the local council has had no other choice than to give consideration to their off-grid plans.

LOCAL CONTACTS

When setting up off-grid systems in a rural community (especially high-profile systems such as a wind turbine or a biogas plant) it is always a good idea to search out local contacts and see what they have to offer. For example, we know a man who sells wood, another who has unlimited mountains of wood chips, yet another who would like to find a home for his manure, a local diner owner who is having problems getting rid of his used deep-frying fat, and best of all a man who knows everything there is to know about fabricating metal.

NOISE, HEALTH, AND SAFETY

You must make sure that you cover all noise, health, and safety issues. Is your turbine going to be noisy? Is your rooftop solar collector going to bounce light into a neighbor's bedroom? Is your project going to spoil the skyline? Consider everything carefully.

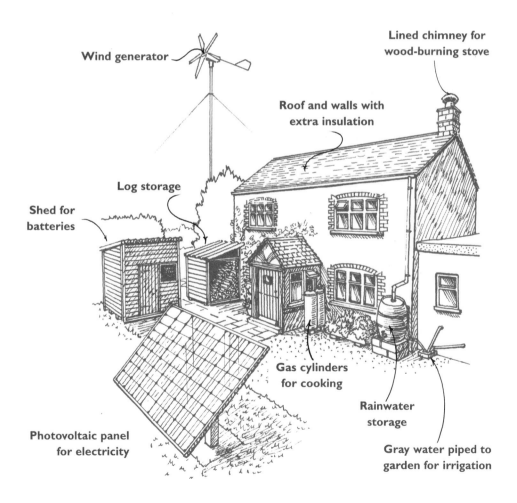

Wind generator

Lined chimney for wood-burning stove

Roof and walls with extra insulation

Log storage

Shed for batteries

Gas cylinders for cooking

Rainwater storage

Photovoltaic panel for electricity

Gray water piped to garden for irrigation

This scenario provides a wide range of options

Rural scenario

The wonderful thing about a rural scenario is that most of the off-grid options are open to you. A cabin in the woods, a farmhouse, a huge house in the middle of an estate, a ranch in rural Montana—the fact that these properties are isolated gives you freedom. Of course, you might not be living in an off-grid dwelling by preference, but having no choice in your peaceful sticks-and-stones cabin other than to come up with an off-grid solution to your energy needs might in itself be liberating.

WITH OR WITHOUT MONEY?

If you have the property and plenty of money, the chances are that you can do it all. It is not so easy if you have the property but only limited funds. When we first moved to the country, we lived in an isolated ruin with no utility services and no road, but somehow we made it work. If you are young, clever, and are thinking about moving into the woods, talk your plans over with your family; if you go abroad, leave funds in your home country, make sure that all your friends and family know what you are doing and why, and then set off on the adventure.

ISOLATION AND TRANSPORT

A rural property can be something of a problem when it comes to bringing in materials and equipment. It is even worse if you are planning to live in a remote area where the terrain is wooded or rocky or if the place is on an island surrounded by water. For example, when we lived in an isolated cliffside community, getting in concrete involved small trucks, little vans, and sacks being heaved in on men's backs or brought around by boat. Transport needs careful consideration if you are thinking about an isolated location.

ISOLATION AND SAFETY

Isolation is wonderful, but only if you are fit and healthy. It is not so good if you develop toothache, your children need to see a doctor, or you have an accident. You must have at least one helper and a reliable line of communication.

SITES OF SPECIAL SCIENTIFIC INTEREST

If you are in a rural location and are planning a large project that involves major ground work (such as digging a lake or cutting down large, established trees), you must make sure you are not living in an area of specially designated interest, meaning that there are protected species of either flora or fauna. Hefty fines can be imposed if protected species are threatened.

RURAL HISTORICAL AREAS

Don't think that if you are moving to the woods you can just dig holes or put up a wind turbine, because it might be that your house and land are in an area of historical significance, such as a local, state, or national landmark. If your off-grid plans are of primary importance and you have a property in mind, then check this possibility before you buy.

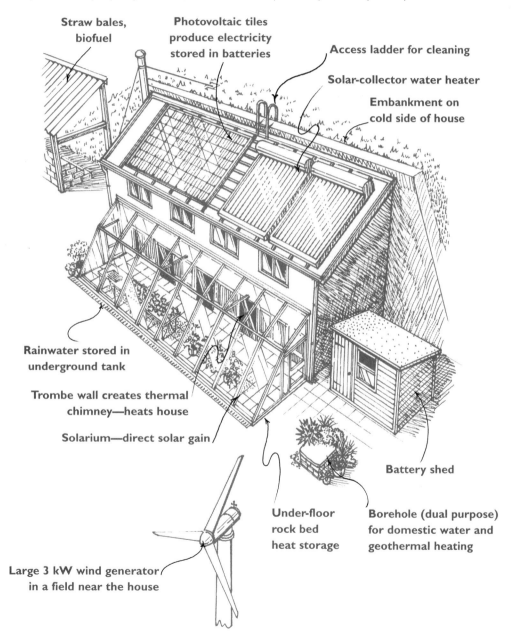

Straw bales, biofuel

Photovoltaic tiles produce electricity stored in batteries

Access ladder for cleaning

Solar-collector water heater

Embankment on cold side of house

Rainwater stored in underground tank

Trombe wall creates thermal chimney—heats house

Solarium—direct solar gain

Battery shed

Large 3 kW wind generator in a field near the house

Under-floor rock bed heat storage

Borehole (dual purpose) for domestic water and geothermal heating

Isolated off-grid options

DOING THE WORK

An important consideration is whether you are going to do all or part of the installation yourself, or whether you are going to pay someone else to do it. Doing it yourself will save money—on the condition that you must do it properly—but it is not always a question of money. Some people simply enjoy becoming involved, digging holes, mixing concrete, and so on. Other people may be so far out in the country, or so short of funds, that they have no choice but to do the work themselves.

THINKING IT THROUGH

- **ARE YOU PHYSICALLY CAPABLE?**
  It is no good rushing all over your property digging holes or removing fences if later on you are going to run out of steam. So, if you are really determined to do the major part of the project yourself, go to your doctor for a check-up before you start. Also think about whether you have the time to do the work, either all in one go or in stages.

- **KNOW YOUR SKILLS AND LIMITATIONS**
  You may be confident that you can perform a certain task, but are you competent? Sometimes the literature that comes with the various systems is so compelling and smoothly reassuring that you can easily be fooled into thinking that you can do everything yourself. This is not always the best idea; for example, with electrical installations it is usually best to pay a specialist to do the work safely and correctly.

- **BASIC RESEARCH**
  Every minute spent finding out about the various options will be money in the bank. For example, when we were first looking to buy our wind turbine, I was amazed at the different prices. After much concentrated research, we got the turbine for a quarter of the first figure quoted. You can save a lot of money by tracing the systems back to source and then "wheeling and dealing."

THINKING IT THROUGH (CONTINUED)

● **CONSIDER THE OPTIONS**

If you want a certain system and you are lucky enough to have lots of money, then you can simply pay up and get someone else to do the work for you. If your funds are limited, you have to find a way of working within those limitations. One way is to take a relevant class and then to build a system or structure from raw materials.

● **DRAW UP A MASTER PLAN**

You can save time and money by being organized and by having a master plan. Start by talking the various options through, and then draw up plans complete with lists and schedules of work. As you do this, you will find that you need to do things in a certain order and that some of your preferred options can be rationalized or amalgamated.

● **SKILL SHARING**

When building, say, a geothermal system yourself, a very good method is to search on the internet, or in your neighborhood, and see if you can swap your skills. You maybe can do the plumbing and fit the heat exchanger, and you know about the pipes and manifolds and heat-sealing the joints, but you know nothing about the complexities of the groundwork—the trenches and holes. Once you have decided what you can't do, search out someone who can do it and who needs a skill you can provide, and arrange to swap skills.

Solar energy

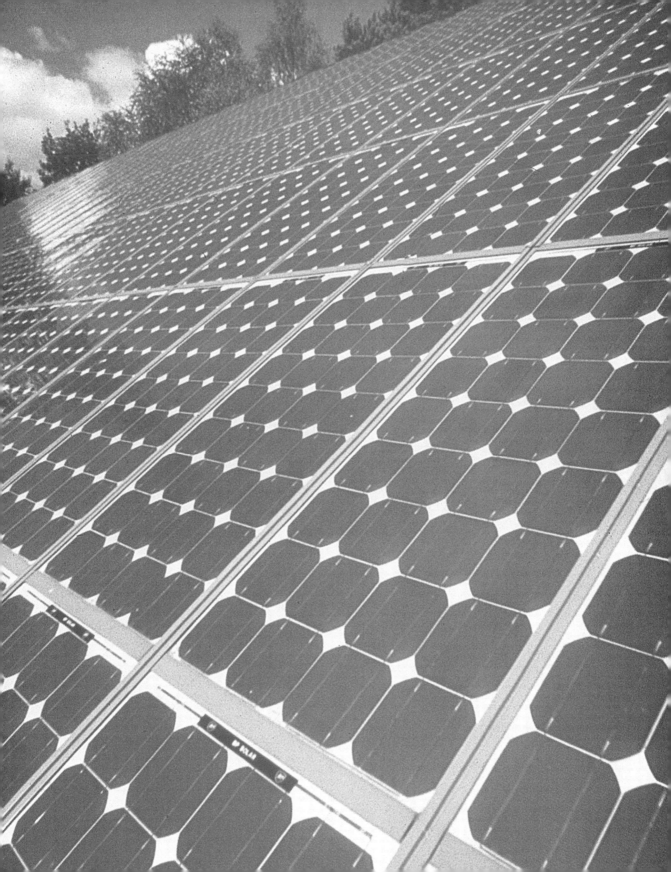

BASICS

Solar basics

"Solar energy" is the term that we use to describe the energy that we get from the sun. There are many ways of harnessing solar energy, so you should carefully consider the various options described below and choose a system that best suits your needs.

PHOTOVOLTAIC CELLS

This system directly converts the light from the sun into DC (direct current) electricity. Each of the cells contains a back contact, two siliconee layers, an antireflection coating, and a contact grid. A batch of cells is able, via an inverter, to create a very useful amount of AC (alternating current) power.

FLAT COLLECTORS

An array of radiator-like metal pipes or glass tubes is set against a heat-absorbing sheet and contained within a glass-front, insulated box. The sun shines on the tubes, and the resultant hot water within the pipes or tubes is pumped through to a heat exchanger in a water-storage tank.

WIDE-ANGLE COLLECTORS

An array of copper pipes or tubes is contained within a curved metal housing. The curved surface maximizes efficiency by focusing the sun through to both sides of a double-sided absorber plate. It is used to heat water.

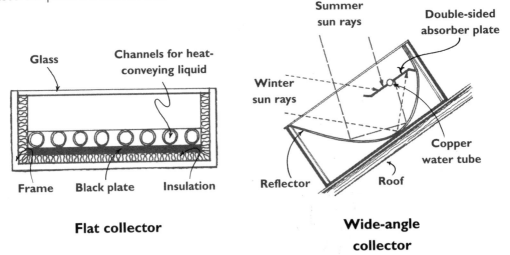

Flat collector

Wide-angle collector

EVACUATED TUBES

This consists of a series of transparent glass tubes, each containing an inner and outer tube, a heat-absorbing surface, a mirror heat-reflecting surface, and a copper heating pipe. The sun's heat is absorbed by a coating on the inner surface of the glass; prevented from escaping by the reflective surface, the heat passes to the tip of the heating pipe, and a copper manifold transfers the heat to a storage tank.

THERMOSYPHON VACUUM TUBES

This is the short name for "coaxial multivalve thermosyphon vacuum-tube water heaters." The system involves a series of glass vacuum tubes linked by valves to a large insulated storage tank. The sun heats the water within each tube, causing it to rise into an integrated storage tank.

PASSIVE HEATING (TROMBE WALLS)

With a typical "Trombe" wall or a Trombe-inspired system (such as a greenhouse or solarium), the sun shines through a sheet of glass to heat masonry walls and floors, with the effect that the space between the glazing and the wall becomes a thermal chimney. Vents in both the glass and the walls of the house are opened and closed as necessary so that the rising currents of hot air within the thermal chimney are directed in or out of the building. This system can be used in conjunction with underground hot rock storage. A Trombe system can be used both to heat and to cool the building.

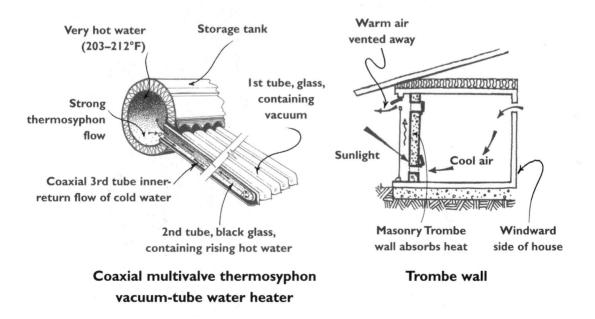

Coaxial multivalve thermosyphon
vacuum-tube water heater

Trombe wall

Advantages and disadvantages

There are so many options that you will need to stand back and consider the pros and cons of the various systems in light of your specific location and your particular energy requirements. You may find that you need to mix-and-match the systems to find the recipe that suits you best.

COMPARING SOLAR OPTIONS

- **PHOTOVOLTAIC CELLS**
  Advantages: These are the only readily available solar option that produces electricity. This is a great system if you want to operate lights and small pumps. They also look good, which is perhaps an important issue when it comes to seeking city or county permits.
  Disadvantages: They are relatively expensive, but prices are rapidly coming down, especially for those made in China (although remember the consequences of this for your carbon footprint). Although the manufacturers claim that the PV cells have a life of 30 to 35 years, they have no data to back up that claim.

- **FLAT COLLECTORS**
  Advantages: Collectors with copper pipes can be obtained as kits, which are good options if you are into DIY.
  Disadvantages: These are now old-fashioned and less efficient than other solar options.

- **WIDE-ANGLE COLLECTORS**
  Advantages: A good option for flat roofs.
  Disadvantages: They are not readily available everywhere.

- **EVACUATED TUBES**
  Advantages: They look good, they perform well, and they have proven successful.
  Disadvantages: These are often sold door-to-door at high prices. You must do your research before parting with your money.

COMPARING SOLAR OPTIONS (CONTINUED)

● **THERMOSYPHON VACUUM TUBES**

Advantages: Top-quality multivalve thermosyphon vacuum-tube solar water heaters (the ones with the large, visible water tanks) are extremely efficient. They are also suitable for all angles of inclination down to 5 degrees, which is almost flat. The system works on a cloudy day and only uses water—there is no need for chemicals. Microsolar water heaters can be installed without the need for electricity or additional water tanks.

Disadvantages: The water heaters with the large, visible water tanks are perhaps a little bulky for most small, neat domestic setups, and might be too heavy for some roofs. This might need to be checked out by a structural engineer.

● **PASSIVE HEATING**

Advantages: Trombe walls and Trombe-inspired setups are a good option if you already have a solarium. They are also good if you are doing new construction and are able to angle your building so that it uses the sun to the best advantage. Combined with rock storage, this is a very good option for a desert climate where it is very hot during the day but cold at night.

Disadvantages: These setups are currently either undervalued or ignored.

Off-grid solar options

What will it power?

Of all the off-grid energy options, solar energy is the most readily available, the most flexible, and the most proven.

PHOTOVOLTAIC CELLS

Photovoltaic panels are a great option for any house (not just those in sunny climates). They will power lights and small appliances but are unlikely to generate enough electricity to run larger appliances like a kitchen stove. Photovoltaics have a lifespan of 25-plus years, they have no moving parts, and they can be "bolted on" to an existing heating or electric system. They are also a good option for a rural vacation home when you might only need small amounts of electricity; a small panel linked to a bank of batteries may give you all the power you need.

FLAT COLLECTORS

These are a workable option for DIY enthusiasts who are looking to build a simple, low-tech system for warming domestic running water using found materials, such as old, flat central-heating radiators or old copper piping. These are unlikely to provide "piping hot" water.

WIDE-ANGLE COLLECTORS

These are more efficient versions of flat collectors and will also warm water.

EVACUATED TUBES

This system also heats water, but it is more efficient than a flat or wide-angle collector. It is a good choice if you live in a city and want low visual impact.

THERMOSYPHON VACUUM TUBES

A more sophisticated version of the evacuated-tube system that will heat water to a far higher temperature than other options. These are more expensive than the alternatives and also include a bulky storage tank on the roof so some authorities may not allow these to be used. The figures suggest that a medium-sized system will provide enough hot water for showers and kitchen use.

PASSIVE HEATING

A conservatory linked to a rock-storage pit is ideal if you live in a sunny climate, favor a traditional approach, and want to do no more than heat space. Hot air is pumped down into the rock storage during the day, and hot air is pumped out during the night. A Trombe-wall system is also a good option if you favor a low-tech, hands-on approach.

SPECIFIC SOLUTIONS

- If you live in a hot climate, you could go for a mix of photovoltaic for lighting, a multivalve system for hot water, and a passive Trombe-wall-type system for daytime cooling and nighttime heating.
- If you live in a city and want a modest system, you could install a vacuum-tube panel to help lower the cost of your existing water heating, plus a passive Trombe solarium to complement your central heating.
- If you live almost entirely off-grid, you could go whole hog and install a massive photovoltaic system for lighting and power, a large multivalve coaxial vacuum system for hot water, and a Trombe wall for space heating.
- If you want lots of hot water, you can't beat a good-sized multivalve vacuum system.
- If you want to cool your home, you could have vents at the floor and roof levels so the house becomes a thermal chimney and expels hot air.

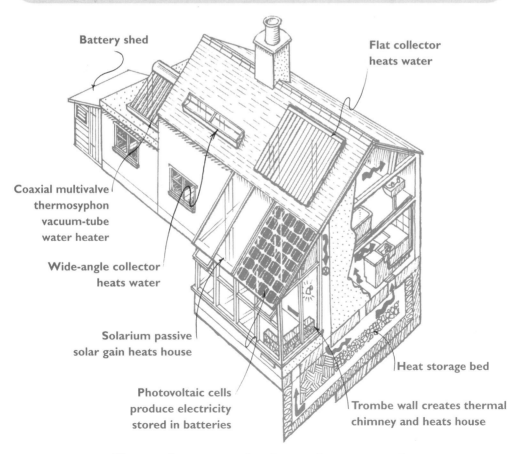

Battery shed

Flat collector heats water

Coaxial multivalve thermosyphon vacuum-tube water heater

Wide-angle collector heats water

Solarium passive solar gain heats house

Photovoltaic cells produce electricity stored in batteries

Heat storage bed

Trombe wall creates thermal chimney and heats house

Choose the system that best suits your needs

SCENARIOS

Townhouse scenario

You live in small townhouse, and you have decided that you want to heat part of your water usage using a vacuum-tube panel and heat some of your living space using a basic Trombe-inspired passive-heat system.

QUESTIONS TO ASK YOURSELF

- Is your house in an area of historical significance, such as a local, state, or national landmark, meaning that you will be limited to using traditional materials? If you are, and there is likely to be opposition to you having a system mounted on the front roof, does the back roof offer possibilities—does it look toward the sun at midday?

- How much of your total hot-water needs are you aiming to cover? Are you just trying to preheat the water to lower overall costs, or are you trying to do it all?

- With regard to a solarium and passive heating, are you able to spend time manually opening and closing vents?

- The solarium needs to have lots of masonry mass and dark surfaces—are you happy with this?

PROPERTY SIZE AND LOCATION

If you live in a house in an area of historical significance where the authorities will want you to use traditional materials, colors, and forms, there might be restrictions that will prevent you from putting solar panels on the roof. It is also no good installing a solar panel if it is going to be grossly overshadowed by another building. Look at the way your property relates to the sun and to neighboring houses before you choose a system.

ORIENTATION

For best results, the roof panel and the conservatory need to be placed so that they face the sun at midday. If the orientation and structure of the roof make it impossible to install a solar panel, you will have to choose another option.

THE STRUCTURE

Carefully consider the roof in terms of structure and materials. If you share the roof with a neighbor, you need to find out if there are restrictions that limit what you can do. Are the roof beams strong enough to take the weight of the planned system? Is the roof covering (felt, copper, steel, tiles, wooden shingles, slates) going to pose difficulties?

THE PHOTOVOLTAIC PANEL

Photovoltaic cells directly convert sunlight into DC electricity. A batch of cells is able, via an inverter, to create a very useful amount of AC power. Assess the size of the system you need by balancing your energy consumption against sunshine and battery capacity. A small photo-voltaic system will operate lights, a computer, and small appliances—all DC. If you want to continue to use your existing AC appliances, you will have to back the system up with an AC inverter.

CONCLUSION

With a townhouse, everything hinges on the neighbors and authorities. The only sensible strategy is to start by talking to interested parties, and then, in light of their needs and concerns, try to get them on your side. This can be very difficult, in as much as the greater the number of people involved, the greater the likelihood that there will be opposition to your proposed system. But it could just be that once your community knows about your off-grid plans and aspirations, you will be able to join forces with a like-minded group and generally cut costs, share the workload, and increase the efficiency of the systems.

Solar options for a townhouse

Hot-climate scenario

You live in a hot climate (somewhere like Arizona), the house is set on a wide-open piece of land, and you want a Trombe wall plus rock storage for winter heating, wide-angle collectors for hot water, and a thermal chimney for natural ventilation.

QUESTIONS TO ASK YOURSELF

- Is your climate extreme—variously very hot and very cold—to the extent that summer cooling is as important as winter heating?

- Does the roof look toward the sun at midday?

- Are you prepared to do structural work to the house?

- How much of your total hot-water needs are you aiming to cover? Are you just trying to preheat the water so as to lower overall costs, or are you trying to do it all?

- Are you able to spend time manually opening and closing vents and shutters?

PROPERTY SIZE AND LOCATION

With this scenario, I am assuming that the house is set on its own grounds and that you are primarily interested in natural, manually operated, passive systems, where cooling is the primary design consideration. You want hot water, and you want to heat the interior, but cooling is most important.

ORIENTATION

What many new off-gridders fail to understand is that the passive heating systems are needed in order to set the cooling systems in motion. So, for example, it is the thermal chimney created by the Trombe wall that increases airflow to the point where cool air is drawn into the house to drive the hot air out. For best results, the main area of roof and the Trombe wall need to be placed so that they look to the sun at midday—facing south to southwest in the Northern Hemisphere, and north to northeast in the Southern Hemisphere. For optimum cooling, there must be lots of vegetation, trellises, ponds, and other features on the east and west sides.

Passive solar-power options

THE STRUCTURE

The building needs to be low, with thick walls, a low-pitched or flat roof, small windows to the east and west sides, and vents and thermal chimneys set high on the roof.

WIDE-ANGLE, NON-TRACKING COLLECTORS

These collectors are uniquely suited to this scenario in that the optics are designed to focus the sun's energy on both the front and back surfaces of the absorber plate. Once in place on the roof, the reflectors can be set so that they concentrate the sunlight to best effect.

THE TROMBE WALL

The Trombe wall is a uniquely exquisite system in that it can be used to provide both heating and cooling. In the summer hot-day/hot-night cooling mode, the hot air within the thermal chimney rises so that cool air is drawn through vents set low on the shaded walls—north in the Northern Hemisphere, south in the Southern Hemisphere. In the winter hot-day/cold-night heating mode, the hot air within the chimney is drawn down into a rock bed and released during the night.

CONCLUSION

With a house set in a desert climate where there are extremes—very hot in the day and cold at night—solar systems present the most cost-effective means of heating and cooling. That said, while passive solar systems (Trombe walls, thick walls, small windows, and roof vents) are incredibly efficient with low running costs, and while new-construction costs are negligible, building such systems into an existing house can be expensive and will be disruptive.

Country-cabin scenario

You have a weekend retreat—a completely off-grid small cottage or cabin—and you have decided that you want a thermosyphon vacuum-tube system for heating water, photovoltaic cells for lighting, and a passive-heating solarium for heating.

QUESTIONS TO ASK YOURSELF

- How much water do you want to heat?

- Do you want to store the water in a cylinder inside the house, or do you want a heater with its own integrated storage cylinder?

- How is your property orientated to the sun—does one side of the roof look to the sun at midday?

- Are you able to do part or all of the work yourself?

- Do you want a completely self-contained water heater with no pumps or electrics?

- Do you have space on the roof for a storage tank and space in the house for a hot-water storage cylinder?

- Are you eventually going to want to live in the house full-time?

PROPERTY SIZE AND LOCATION

More than anything else, the property needs uninterrupted sunlight. With the ideal layout being a house that presents the systems directly to the sun at midday, you need to look at your cabin or cottage and see if it is going to work out.

ORIENTATION

Ideally, the roof-mounted systems and the solarium need to look to the sun at midday. The advantage with most country properties is that the size and orientation are flexible.

THE STRUCTURE

The roof must be structurally sound and therefore strong enough to take the additional weight of the systems.

THE PHOTOVOLTAIC PANEL

A modest system—big enough to power a weekend cabin—is equal in cost to, say, a small generator setup complete with batteries and oil lamps. A small photovoltaic system would, with careful management, just about operate lights, a water pump, a computer, and a few small appliances—all DC. For AC power you would need to back the system up with an AC inverter.

THE THERMOSYPHON VACUUM TUBES

Microsolar water heaters are unique in that they can operate in very low temperatures without freezing, without chemicals, and without any form of backup heating. They can work on damp, overcast days when other vacuum-type heaters come to a halt. They are a good option for this scenario in that they are self-contained—no electrics, no extra tanks, and minimal pipe work. They can be mounted on a flat or pitched roof, as long as they are angled up so that they are looking at the sun at midday—meaning at least 5 degrees off horizontal. In most situations, this system can be plumbed directly off the main water line.

PASSIVE HEATING

With a typical Trombe wall or a Trombe-inspired system—as with a greenhouse or solarium—when the sun comes up, it shines through the glazing to heat masonry walls and floors, and when the sun goes down, the heat stored within the structure of the cabin is given off to warm the interior space.

CONCLUSION

The wonderful thing about a small off-grid cabin in the country is that there is plenty of room for you to do your thing. If you want to add a system here or change a detail there, then you can, to a great extent, just get on with it. You can make lots of noise, do the work in the early hours, or have ramshackle structures—in the country there is no one to care or complain.

CASHING IN ON YOUR SOLAR POWER NOW!

Thanks to the Emergency Economic Stabilization Act, every U.S. homeowner can find incentives for generating their own energy through solar systems. Tax rebates on city, state, and federal levels can return up to 30 percent of the cost of the panels and installation as long as your system is up and running within an allotted time. Check with your state and local governments before purchasing your system to ensure you get your due credits. See the resources on page 188 for more information.

SYSTEMS

A passive-heating system

If you live in a small house in the country and you are prepared for an extensive remodel, or if you are in the first stages of designing a new building, here is how to install a Trombe-wall system complete with rock storage so that you can heat the house.

WHAT YOU NEED TO CONSIDER

Passive solar heating is a great low-cost option in new buildings—a structure like a Trombe wall will hardly figure in the financial scheme. If you are going to rework an existing property, however, not only will the initial costs be high but you will have to take into account the disruption and the need to excavate a sizable hole. When building a rock-storage system on an existing property, there are three options: (1) build it under an existing floor, in which case you will have to think about mess, stresses to the structure, and the implications of breaking through the floor and underlying damp-proof course (DPC); (2) build it close up against the outside of the house, in which case you will have less mess but the pit will have to be waterproofed and superinsulated; or (3) perhaps best of all, build it under a new addition.

DIRECT SOLAR GAIN

If you build a basic greenhouse-type solarium against the house, the walls and floors warm up during the day and give off heat at night. The bigger the solarium and the larger and darker the storage mass—the walls and floors—the greater the capacity to store heat.

THE TROMBE WALL

At its simplest, a Trombe wall is a dark-painted brick or concrete wall fronted with glass, with a small space between the glass and the wall. When the sun shines through the glass, the wall gets hot, with the effect that the air in the space—in the thermal chimney—rises. If vents are built into the structure at the top and bottom, it is possible to variously heat and cool adjoining spaces.

LOCATION AND BUILDING SHAPE

The location and the way the systems are presented to the sun are vital. If the site is heavily wooded with tall trees, overshadowed by buildings or mountains, or positioned so that the roof faces away from the midday sun, overall efficiency will suffer. The ideal is to have the building aligned on an east–west axis, with the windows/glass/Trombe systems facing the sun at midday, and the main roof angling down away from the sun.

HEAT STORAGE

Heat can be stored in masonry structures such as brick and stone walls, concrete floors, water walls (glass or plastic tanks full of dark water), well-insulated tanks of water, ponds, and rock bins or beds. When the sun goes down, the heat is either simply given off or pushed by means of a small fan from, say, a rock-storage bed or a greenhouse to various remote parts of the house. You could, for example, transfer heat from a greenhouse to a dark, cold room on the east side of the house.

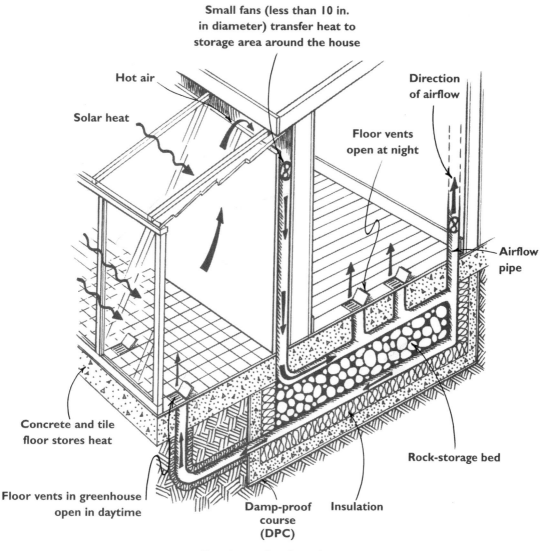

Small fans (less than 10 in. in diameter) transfer heat to storage area around the house

Hot air

Solar heat

Direction of airflow

Floor vents open at night

Airflow pipe

Concrete and tile floor stores heat

Rock-storage bed

Floor vents in greenhouse open in daytime

Damp-proof course (DPC)

Insulation

Passive solar heating

A natural-cooling system

This is an ideal system if you live in a very hot region in a rural setting and you want to cool your house using natural, passive, low-cost means, and as much as possible you are trying to do away with all electromechanical systems, including pumps and fans.

WHAT YOU NEED TO CONSIDER

In the past, people built small, dark, low-to-the ground cottages and homesteads using natural materials. The walls were thick, the roofs were low, the floors were damp, the windows were small and shuttered, there were porches on the sunny side, and there were lots of shrubs and trees all around. This all helped to keep the house cool, and you are trying to achieve a similar effect here.

LOCATION AND BUILDING SHAPE

The ideal is to have the house set low in its surroundings, with lots of trees, shrubs, shaded patios, and screens on all sides except for the one that faces the sun at midday. If you get it right, the Trombe windows and solarium will face the sun at midday, but the rest of the house will be in heavy shade. Better still, design the yard or outside space so that there are splashing water features mixed in with the shrubs and patios.

MATERIALS

Natural materials are best—brick, stone, ceramic tiles, wood, earth, and fabrics such as cotton, linen, and wool. Interestingly, many of the values, methods, and procedures that are advocated for natural, passive solar heating are adaptable to natural cooling. For example, while dark-tiled masonry floors and walls absorb heat, they do it in a way that holds down sudden rises in temperature. The outside of the building is best painted or treated with light colors that reflect sunlight. That said, you need to cover the ground with vegetation and have shades around windows and doors so that the sunlight is absorbed rather than reflected back into the house.

CONVECTION, VENTILATION, AND AIRFLOW

You need to install large vents high up in the walls, or better yet in the ceilings, and smaller floor-level vents on the coldest and windiest side of the house. If you get it right, when the hot air rises up and out through the ceiling vents, it will drag in cool air at floor level. Better yet, you could build vents and chimneys even higher at roof level so that the air flowing over the roof creates a suction effect that draws hot air up and out of the house.

EVAPORATION

If you have ponds or water features just outside the house on the cold, shaded side, air rushing into the house at floor level will pass across the surface of the water to create an evaporative cooling effect.

COOLING BY MEANS OF A TROMBE WALL

In the hot-day/hot-night scenario, with the top and bottom vents in the window open, and the vents in the wall closed, the rising air in the thermal chimney between the glass and the wall creates a circulating stream of air that helps to keep the temperature within the wall down. If the day and night are very hot, and you open the top vents in the window, the vents at the bottom of the Trombe wall, and, vents on the cold windward side of the house, then the hot air that rushes out of the top window vents will draw in cold air from the far side of the house.

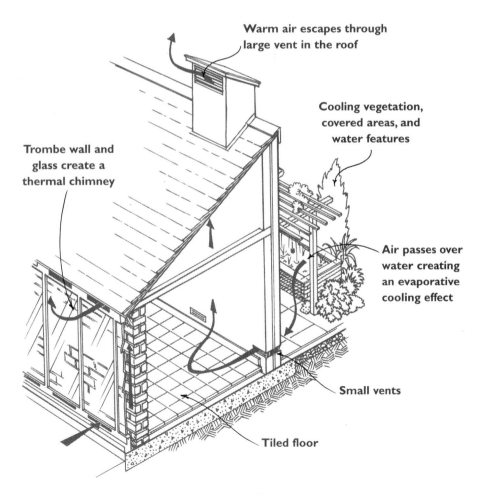

Passive cooling

A DIY flat collector

If you enjoy the idea of solar heating and you want to heat water, but you are on a very tight budget, you might want to try at making a solar collector from low-cost—or, better yet, no-cost—salvaged materials.

HOW IT WORKS

In action, the sun heats the antifreeze mix within the collector, the pump (operated by the photovoltaic panel) pushes the heated antifreeze through the heat the exchanger coil within the storage cylinder that it heats the domestic water.

LOCATION AND ORIENTATION

The ideal orientation is to have the collector mounted on a roof at an angle of between 30 and 50 degrees so that it faces the sun at midday.

MAKING THE COLLECTOR

When you come to making the manifold (the pipe work within the collector) and are wondering about the length, number, and spacing of the risers, the rule of thumb is the bigger the better. In other words, while our drawing shows ten 3 ft. risers spaced about 2 in. apart, there is nothing to say that you can't have twelve at 1 in. apart, or whatever you want. Yet again, while we have used soldered junctions, you might prefer to go for the easier but more expensive option and use spanner-fit compression joints. If you look at the drawings, you will see that, from back to front, the materials to make the box that holds the manifold are: sheet of marine plywood, one 2 in. thick slab of foil-faced foam insulation, fiberglass roof insulation, the ladder of copper pipes complete with its copper fins, and a sheet of glass bedded in silicone. Again, if you want to replace the foil-faced foam slabs with fiberglass topped with aluminum foil, then fine.

The collector plates (the flat plates that fit on the vertical risers that make the manifold) are made from thin-sheet copper cut to fit and soldered in place directly onto the back of the copper pipes. As for the box that contains the riser tubes, plates, and all, you can frame it up using treated wood sections (as shown here), aluminum, strips of plastic, or whatever you have on hand.

INSTALLATION

The finished collector will need to be installed on the roof by a competent carpenter and plumbed and wired into your system so that it complies with local regulations.

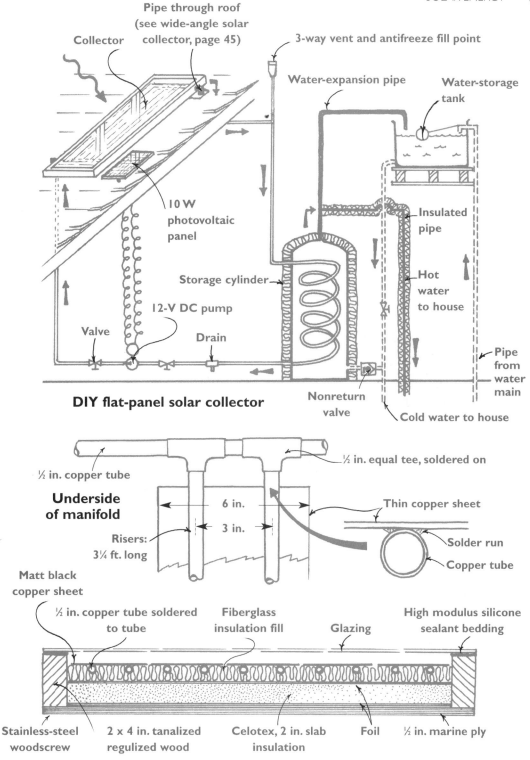

DIY flat-panel solar collector

Collector

Pipe through roof
(see wide-angle solar
collector, page 45)

3-way vent and antifreeze fill point

Water-expansion pipe

Water-storage
tank

10 W
photovoltaic
panel

Insulated
pipe

Storage cylinder

Hot
water
to house

12-V DC pump

Valve

Drain

Pipe
from
water
main

Nonreturn
valve

Cold water to house

**Underside
of manifold**

½ in. copper tube

½ in. equal tee, soldered on

6 in.

3 in.

Thin copper sheet

Risers:
3¼ ft. long

Solder run

Copper tube

Matt black
copper sheet

½ in. copper tube soldered
to tube

Fiberglass
insulation fill

Glazing

High modulus silicone
sealant bedding

Stainless-steel
woodscrew

2 x 4 in. tanalized
regulized wood

Celotex, 2 in. slab
insulation

Foil

½ in. marine ply

Cross-section through collector

A wide-angle collector

If you live in a very open area of a warm country—somewhere like the desert regions of the U.S., Australia, or Spain—and your house has a large area of flat roof, you might decide that you want an array of wide-angle collectors to heat all your domestic water.

WHAT YOU NEED TO CONSIDER

Wide-angle collectors, sometimes called "concentrator collectors" or even "nontracking wide-angle concentrators," are uniquely interesting in that their structure is such that they focus the sun's energy on both the front and the rear surfaces of the absorber plate, which reduces the initial cost of the system. The absorber plate, coated on both the front and rear surfaces with a matte black finish and backed with an absorber, is angled within a curved reflector housing in such a way that the collector is able to absorb both direct and diffused sunlight over the whole aperture. A bonus of this system is that while most traditional collectors use a multijointed ladder-grid of water-filled copper tubes—a structure that is inherently faulty because there are so many soldered joints—this system is designed so there are no joints within the collector housing itself. So, if a leak occurs it does so on the outside rather the inside of the curved housing. While wide-angle collectors can be bulky in size, heavy in weight, and a little of an intrusive eyesore, all such factors are relatively insignificant in the context of the systems being mounted on flat roofs in areas that are well away from urban-planning sensitivities.

LOCATION, BUILDING SHAPE, AND MANUAL ADJUSTMENT

As with any other roof-mounted solar system, the way that the system is presented to the sun is all-important. If the collectors are overshadowed by buildings or trees, the overall efficiency of the system will suffer. The ideal is to have the collectors arrayed on a flat roof with the reflectors angled so that they face the sun at midday. With wide-angle collectors, bearing in mind that the amount that the curved reflectors can be moved within their boxes is restricted, the boxes need to be set at the optimum angle for the specific system. As a general rule of thumb, the angle should be approximately equal to the latitude plus 15 degrees.

With a small domestic system, get your system up on the roof, point the individual reflectors at what you know to be the position of the sun at midday, and then make any necessary adjustments. There are three practical options for adjusting the angle: You could change the angle from month to month; you could fix the array at what you consider to be the best optimum year-round angle; or you could make semiannual adjustments for winter and summer sun.

TRACKING ARRAY FRAMES

If you want to achieve maximum efficiency—and this would be a good option if you can afford to install a photovoltaic system to provide a small amount of electricity—there are electrical systems that allow you to install tracking devices that will constantly move the frames so that they follow the sun. If you are unable to go clambering up on the roof or you are simply too busy, then an automatic motor-controlled tracking system is a good option. Trackers can be expensive, and they do need to be maintained at regular intervals, but they will allow you to run the system at maximum efficiency.

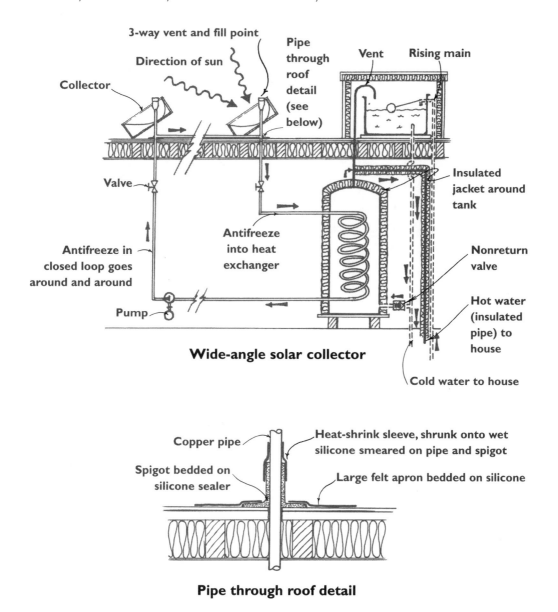

Wide-angle solar collector

Pipe through roof detail

A thermosyphon vacuum-tube system

This type of system is ideal if you live in a small house in a rural setting and you want to use a solar water-heating system—you want a low-maintenance, standalone setup that you can install without making too many changes to existing installations.

WHAT YOU NEED TO CONSIDER

Currently, the "microsolar" system is so far ahead of other apparently similar systems that it is vital that you are able to tell it apart from its various clones. The difference is not so much in appearance—because there are superficial similarities—but in the technical details. For example, while lots of vacuum solar water heaters of this type look outwardly much the same—they have an array of glass tubes topped by a large cylindrical water-storage tank—the genuine microsolar system has a patented coaxial multivalve arrangement that ensures that there is no premature mixing between the hot and cold water within the tubes, no overheating, no loss of efficiency on cloudy days, and no extremes when it comes to the range of setup angles. A genuine microsolar system will work efficiently at a whole range of inclinations—from 90 degrees right down to 5 degrees to the horizon. What this means is that this system can be mounted on just about anything from a vertical wall to a flat roof.

This system is able to deliver hot water at 105°F on overcast, rainy days, 120–140°F on averagely sunny days, and 140–175°F on hot sunny days. Being mindful that the system has its own integrated water-storage tank, so you don't need an additional tank in the house, you need to figure out which size you want—58 gallons or 78 gallons.

LOCATION AND BUILDING SHAPE

The joy of this system is its flexibility. It is good if you have a pitched roof that more or less looks to the sun at midday, but it can be mounted just as well on a flat roof.

THE PLUMBING

One of the big bonus points for people who are starting from scratch, or trying to "plug in" to an existing system, is the fact that this system is self-contained with its own integrated water-storage tank. It is also very straightforward to install. You need to mount the system on the roof, run a cold-water feed from the main water line, run the hot water down to the various hot-water taps around the house, heavily insulate all the pipes—especially those that rise above the roof line—and make sure that the through-roof pipe is protected by a spigot and sleeve.

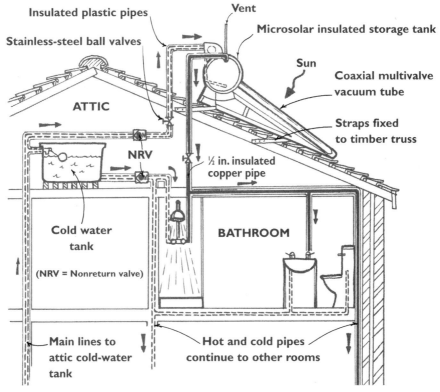

Insulated plastic pipes

Vent

Microsolar insulated storage tank

Stainless-steel ball valves

Sun

Coaxial multivalve
vacuum tube

ATTIC

Straps fixed
to timber truss

NRV

½ in. insulated
copper pipe

Cold water
tank

BATHROOM

(NRV = Nonreturn valve)

Main lines to
attic cold-water
tank

Hot and cold pipes
continue to other rooms

**Microsolar collector
thermosyphon-solar-water-heater installation**

IMPORTANT NOTES

- You must make sure that you are legally able to link directly to the water main.

- The microsolar system is pressurized—the outlet pressure will be approximately equal to inlet pressure minus 15 percent (for pipe friction).

- For safety's sake, you must install mixer taps in the shower.

- Short, well-insulated pipe-runs equate with high efficiency; long, uninsulated pipe-runs will result in disappointing temperatures.

- As with most solar systems, you will have to modify your needs slightly. So, you might have to cut back on hot-water usage if there is a long spell of overcast weather, but when the weather is hot there will be an unlimited supply of hot water.

A small solar photovoltaic system

If you live in an isolated cabin-type property with no electricity or gas, you could install a basic photovoltaic system to power lights, a small water pump, a computer, and small appliances such as a TV and a radio.

WHAT YOU NEED TO CONSIDER

With a photovoltaic system, sunlight goes in at one end and electricity comes out at the other, and is stored in a bank of batteries. A typical photovoltaic system is made up of an array of photovoltaic panels, a battery charge controller, a DC switchboard, a bank of batteries, an inverter, an AC switchboard, and a range of cable, wires, switches, and cutouts. There are many different systems to choose from—everything from a neat 3 x 150 W solar panel setup that will power a small weekend cabin to a 3.8 kW setup that will power a four-person house. Research suggests that a typical standalone, off-grid family home, with a backup of a wind turbine and/or a small generator, needs a 1.2 kW array, but a small, frugal home could get by with, say, a 500 W system. The panels need to be mounted either on the roof or on a free-standing array frame, and they should ideally be positioned facing toward the midday sun and sheltered from strong winds.

THE BATTERY BANK

A 3 x 150 W cabin setup will need a bank of 4 x 350 A.h. deep-cycle batteries, while a family-sized 1.2 kW setup needs a bank of 8 x 1080 A.h. batteries. The more batteries you have, the larger your storage capacity.

THE CHARGE CONTROLLER

The charge controller regulates the electricity as it comes from the photovoltaic panel. The controller ensures that the batteries are always presented with the correct voltage/current. The controller reads the state of the batteries and channels the charge accordingly. Think of the controller as a control portal between the panels and the batteries.

THE INVERTER

The inverter changes the DC electricity as it comes from the battery bank into standard AC electricity—the type of electricity that most of us use to power our homes. The size, shape. and power of your inverter will relate to the size of your battery bank.

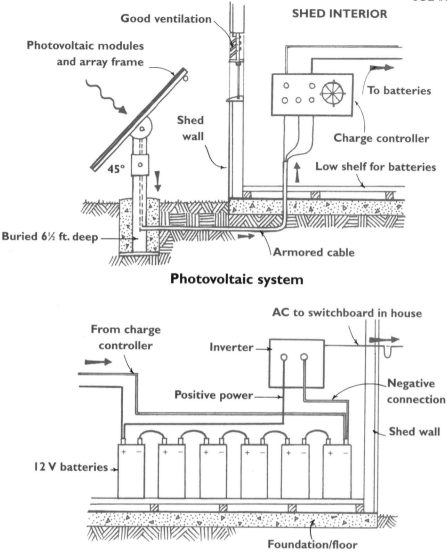

Photovoltaic system

Shed interior continued

THE DC SWITCHBOARD

This is a box that houses all the meters and circuit breakers that control the DC system. The DC electricity goes from the array, through the controller, and through the DC switchboard.

THE AC SWITCHBOARD

For safety's sake and for easy reading, the AC switchboard is best set up as a separate item some short distance away from the DC switchboard. The AC switchboard needs to have all the usual circuit breakers, safety switches, cutouts, and leakage breakers—just as with the usual domestic-consumer breaker box.

HINTS AND TIPS

The whole area of photovoltaic energy is growing and developing at an amazingly fast rate, so much so that it will soon be one of the world's biggest industries. Because of this, and with solar energy options rapidly going up and prices coming down just as quickly, it is vital that you do your research and understand all the potential pitfalls before committing yourself to any single system.

THINKING IT THROUGH

- **DO YOUR RESEARCH**

  One look at the various solar options will show you that there are so many exciting possibilities out there that it is very difficult to settle on a single system. Then again, no single solar system will do it all—yet. Don't be rushed, do as much research as possible, and always remember that you can usually quarter your costs simply by doing the work yourself. You have little choice at present other than to go for a hybrid system.

- **RECOUPING COSTS**

  In many ways the notion of recouping costs is nonsense. When we buy a stove, we don't think that somehow we have to make the stove pay for itself. So, unless you are starting with a completely clean slate on a new property, you will need to start thinking differently and stop worrying about recouping your costs. The fact that you are changing your everyday living habits for the better is a much more important consideration.

- **INSULATION**

  Perhaps the whole subject of insulation lacks glamour, but good insulation is vital to the whole off-grid energy setup. Why spend lots of money heating and cooling your house if the whole structure is so poorly insulated that it leaks like a sieve? First thing's first: No matter your climate, and no matter which systems you use, you must fully insulate your whole house. The best first step in creating a green, energy-efficient, off-grid system is to start with energy conservation. Research suggests that good insulation, including draft-free windows, insulated curtains, solar porches, and storm doors, will help to cut energy costs by a staggering 50 to 90 percent.

THINKING IT THROUGH (CONTINUED)

- ### SHAPING YOUR SYSTEMS TO SUIT YOUR NEEDS

  Are you going to be able to afford such and such a system? Will you be able to do any part of the work yourself? Is your primary need cooling rather than heating? Is your area predominantly windy, sunny, cold or rainy? Is it a big problem if your hot water, electricity, or heat runs out? Will it matter if you wake up to a cold house? Are you fit and strong enough to carry logs? You must tailor your system to suit your budget, skills, and needs.

- ### SOLAR CHIMNEYS

  A well-insulated Trombe-type wall will dramatically increase your amount of solar-energy gain. If after insulating every cavity, nook, and cranny you can only afford a single low-cost solar system, this is the one to go for.

- ### PHOTOVOLTAIC SOLAR TYPES

  There are single-crystal solar cells, polycrystal solar panels, amorphous silicone panels, multicrystalline silicone panels, copper-indium-diselenide cells, gallium-and-arsenide cells, and no doubt more new forms are being developed right now. The best you can do is keep up with the research so that you can make an informed choice.

- ### ARE PHOTOVOLTAICS THE ANSWER?

  Current thinking in the U.S. and China suggests that there is an outside chance that photovoltaic technology is perhaps the "silver bullet"—the single, big energy solution—that the world is desperately seeking. This could mean the end of the need for massive power plants, and the birth of a world where every house has photovoltaic panels built into its very structure—photovoltaic tiles on the roof, windows with panels built into the glass, each outside light having its own standalone panel. Time, of course, will tell if this turns out to be true.

- ### ELECTRICAL INSTALLATIONS AND SAFETY

  Electrical installations must comply with local regulations. Ask a qualified electrician to install your system. You are also likely to require a safety inspection.

Wind energy

BASICS

Wind basics

Wind energy equates with wind turbines, of which there are several types to choose from. Be warned that, while some turbines are being marketed as rooftop machines, some experts think that a rooftop setup will invariably result in problems with wind turbulence and vibration. Many experts specifically advise against "building-mounted" installations. You could perhaps compromise by having the turbine mounted close to your house at roof level, with the mast running directly down to the ground but supported by the house.

Bad position—wind turbine in sheltered valley

Good position—wind turbine free of obstructions

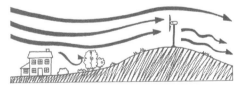

Ideal position—wind turbine on top of hill

ROOF-MOUNTED MINI TURBINE

This turbine is specifically designed to be mounted on the highest point of the roof. The makers of this turbine, with a rotor diameter of about 7 ft. and a wind cut-in speed of about 7½ ft. per second, claim that it is able, depending upon the site, to produce 500–1000 kW of electricity. Bear in mind, however, the advice given above regarding roof-mounted turbines.

SMALL FREE-STANDING TURBINE

This can be any turbine mounted on its own free-standing, guy-supported mast that is able to produce up to 2 kW of electricity. It is small enough to be termed "temporary," and as such should not require permits to install. An average system of this size will have a 12 ft. diameter rotor on a 27–30 ft. high tower.

LARGE FREE-STANDING TURBINE

Any tower-mounted turbine rated as 3 kW or more is a serious piece of equipment. It will, under favorable conditions, power the average small house, but it will probably require permits, careful siting, a ground/engineer survey, and so on.

FREQUENTLY ASKED QUESTIONS

- **Will I need city/county permits?** You almost certainly will if you live in a populated area. If you live in a rural location permits may not be required, but since regulations are constantly changing, it is wise to ask about current standards and restrictions.

- **Can I connect my turbine directly to my electrical circuit?** No, you need a grid-tie inverter to convert the DC electricity generated to a standard AC supply.

- **My house is surrounded by trees—is this a problem?** Though much depends upon wind direction and the precise lie of your land, you will need to mount the turbine on a mast that places it at a higher level than the trees.

- **How can I find out about the wind in my area?** For most developed countries, you can obtain local wind-speed data by searching on the internet.

- **How much noise will a turbine make?** The degree of noise will depend upon specific size and siting. A medium-sized turbine will make a loud swishing noise, but the sound of the turbine will, to a great extent, be cloaked by the noise of the wind.

- **What is the best way of using the power?** You can store it in a battery bank for powering lights and appliances, or you can use it directly for heating rooms and for hot water.

Advantages and disadvantages

A wind turbine can be an exciting and efficient option, but only if you get it right. By its very nature, it is a dynamic structure that must be considered from a whole range of viewpoints, including visual impact, efficiency, maintenance, and cost of the battery bank.

ADVANTAGES

Combination with solar power   A hybrid system of a wind turbine and photovoltaics is a good option in that it nicely covers opposite extremes of the climate spectrum. Windy or sunny, your system will be making energy.

Availability, choice, and cost   In the last few years, choice and availability of wind turbines have gone up and costs have come down. It is possible to import very good, low-cost machines from China.

Iconic and dynamic   For lots of people, a wind turbine is more than a system; it is something against which they measure their commitment to clean, green, off-grid energy.

Seeing is believing   You can see a wind turbine at work; when the wind blows, the sails, props, or blades go around and you can almost feel the energy being made.

Ideal for isolated off-grid use   A large wind turbine is a good option for a truly rural community. Because the traditional electromechanical technology is easily understood, it is relatively easy to maintain.

Electricity   There are only two readily available options for creating electricity— photovoltaics and wind turbines, and at present the latter is the more well-proven option.

Cutting costs   According to research, a good-quality domestic wind turbine produces cheaper electricity over its 20-year lifetime than grid-purchased energy.

DISADVANTAGES

Safety   Wind turbines are, by virtue of their very size and dynamic motion, potentially very dangerous pieces of machinery. If a wind turbine falls down, it could easily kill you. For this reason, it needs to be expertly installed and everything must be completely secure. This is an extremely important issue if you have children and animals or if your site is in any way open to the public.

City/county permits   In the U.S., the U.K., Australia, Europe, and other developed countries there is no certainty that permits will be granted, so always check before you proceed.

Space   A wind turbine needs a good area of land. For example, a small turbine with a tower height of 20 ft., you need a circle of ground with a diameter in excess of 40 ft.

Winter weather   You will need the most power in the winter, so a wind turbine is a poor option in areas where the climate is characterized by fog, frost, and still air.

Windy or not?   Even though it is usually easy enough to find out the average wind speed for your general area, it is not so easy to find out the precise wind speed at a point 20 ft. above your property. A good way of assessing wind speed is to look at the trees and compare their degree of movement against the "Griggs-Putnam Index of Deformity" (available on the internet).

Dangerous wind speeds   If the wind decides to blow your house down, there is nothing you can do about it. Research suggests that your turbine will be more at risk from constant turbulence and buffeting than from high winds. Always go for top quality in your tower, foundations, and guy wires. If you are nervous about high winds, or you live in an area that is known for its gale-force winds, perhaps a wind turbine is a bad idea.

What will it power?

A wind turbine, the most iconic of all the off-grid, renewable energy options, will produce electricity and heat. Be warned, though, that it is a very serious piece of engineering machinery that will constantly need to be monitored, maintained, and managed.

WIND ENERGY USES

- If you live in a rural, windy area, using a turbine with a bank of batteries will enable you to operate everything in your house except the stove, oven, and shower.

- If you take an existing water-storage cylinder complete with an immersion heater, match the heater to your turbine, and then connect your turbine directly to it, it will supply you with domestic hot water.

- A wind turbine can supply heat through a resistive heater (like a storage heater) or it can be fitted with a controller and automatic switch and programmed so that it first recharges your batteries, then heats the water, and finally supplies heat.

USING A TURBINE WITHOUT BATTERIES

A wind turbine without batteries can be used to power resistive loads, meaning items such as immersion and convector heaters that contain a heating element. You must make sure that the total power rating of the heating elements always exceeds the maximum supply potential of the turbine—otherwise on a windy day the heating elements will burn out. A good resistive option is to connect to a water-cylinder immersion element that is always set in the "on" position. This setup will use all available power from the turbine.

USING A TURBINE WITH BATTERIES, CHARGING
REGULATOR, AND AN AC-POWER 240 V SINE-WAVE INVERTER

With this setup, you can power all your appliances that have a motor or transformer, such as a stereo, computer, TV, DVD player, washing machine, or dryer. When designing the system, be aware of two factors: The appliances will not all be on at the same time, and the turbine will only be charging the batteries some of the time. The battery bank must be large enough to take over when the wind turbine is "at rest"—about 70 percent of the time. The turbine itself needs to be large enough to supply all your requirements while only working 30 percent of the time. The turbine needs to be four times larger than your daily usage (see opposite page).

WORKING OUT YOUR POWER NEEDS

Calculate the total kilowatt-hours (kWh) you use (see page 11) and multiply this figure by four, as the wind turbine only works about a quarter of the time; this will give you the value of the wind turbine in kilowatts (kW). If at the end of all these calculations, you simply can't afford a turbine big enough to match your estimated needs, you have three choices:

• Cut your power usage by economizing or reducing the number of appliances you use.

• Supplement your system with another off-grid power source, such as another wind turbine or photovoltaic cells.

• Do both of the above.

Remember, the general rule is the bigger the turbine, the more efficient it is, and the quicker it will pay off its initial capital cost.

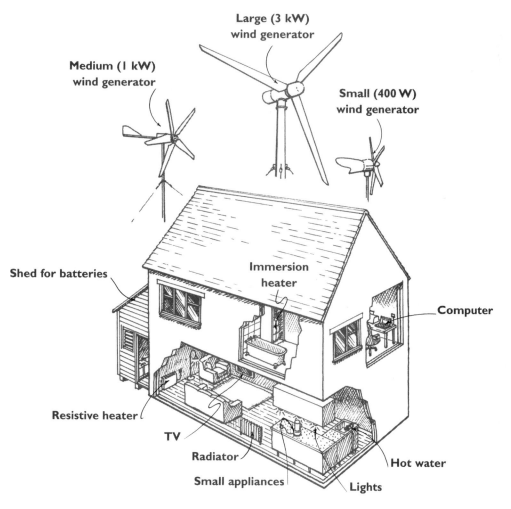

Wind-energy options

TOWNHOUSE SCENARIO

If you live in a good-sized townhouse, you may decide that you want to install a wind turbine to make some part of your own electricity. The idea is that you are going to create a little, dedicated off-grid system for your workshop.

QUESTIONS TO ASK YOURSELF

- Do you need city/county permits?

- Do you have nearby neighbors? If you do, have you asked them for their opinions?

- Have you researched the wind speeds in your area?

- Is the structure of your house—the walls, gables, and so on—strong enough to take the stress?

- Have you researched the total cost (turbine, tower/mounting brackets, controller, inverter, and batteries)?

- Is there enough space for a dedicated battery shed?

PROPERTY SIZE AND LOCATION

As with all off-grid options, the size and location of your house will shape your needs—more so if you live in a densely populated area. In this case, you have no choice other than to go for a small machine.

A TURBINE ON THE ROOF

As already mentioned (see page 54), experts are divided about whether mounting a turbine on the roof is a good idea. Research suggests that while many rooftop machines do work—in the sense that they make electricity and they don't damage the structure or make too much noise—the fact is that the overall performance will, to a greater or lesser extent, be affected by wind turbulence.

A SMALL 400–500 W TURBINE AND NOISE

While most small turbines are smooth running in operation, they do make some noise. For example, although two well-known manufacturers claim that their machines are "virtually silent running," figures suggest that the actual noise levels are somewhere around 35–50 decibels.

To give you some idea of what this means, in a bedroom in the country you will usually hear 35 decibels, in a car traveling at 40 m.p.h. you'll hear 55 decibels, and in a busy office you'll hear 60 decibels.

WILL A ROOF-MOUNTED TURBINE DAMAGE THE HOUSE?

The answer depends upon the size of the turbine, the way it is mounted, and the age and design of the house. For example, some old houses have poorly built, single-brick gable ends. If you look at the illustration below, you will see that a safe method is to mount the turbine as recommended by the manufacturer—complete with well-engineered brackets, antivibration pads, and so on—and then go one step further and run the mast straight down into a ground-level foundation block.

CONCLUSION

Everything depends upon your house and your town, but you will almost certainly have issues with construction permits and codes. The efficiency of the turbine will, more than likely, also be affected by wind turbulence. However, if you install the turbine according to the manufacturer's instructions, the noise will be so low as to be hardly noticeable, and the turbine will produce electricity.

Small turbine on a townhouse

A townhouse wind turbine

If you live in a townhouse, you might want to lower your energy costs by installing a small low-cost wind turbine. The first thing to understand is that the whole procedure is fraught with pitfalls, from local regulations to expensive machines that simply are not up to the task. You simply must do a great deal of detailed research before you embark upon any installation.

OFF-GRID STANDALONE SYSTEMS

The whole idea of an off-grid or standalone system is a wonderfully inspiring notion—just you and the turbine and no more utility bills—but it is not easy. For example, one research program in Australia suggests that to make it all work you need a hybrid system made up of a wind turbine, an array of photovoltaic panels, a house design that uses various passive solar systems, plus a bottled-gas system and a diesel generator. So, while an off-grid system is possible, you must be flexible enough to modify your energy needs to suit what is available.

TIE-IN GRID-CONNECTED SYSTEMS

Grid tie-in is a great idea—you push all your overflow energy back into the grid, and the electric company pays you, or you pay them, the balance between what you give and what you take. The downside is that in some developed countries there are, as yet, no clear-cut mechanisms for exporting electricity back into the grid, although many forward-thinking countries do have very good tie-in systems.

CHOOSING A TURBINE

There are now lots of wind turbines to choose from. That said, and this is particularly important in the context of domestic wind turbines in urban areas, it is vital that your turbine be virtually silent in operation, reliable, visually nonintrusive, and altogether safe. A good choice for a rooftop situation (if this is your only option) is a 250–400 W turbine.

MONITORING SYSTEMS

If you hoist a sail or flag and leave it to blow in the wind, sooner or later a fierce wind will come along and blow it to pieces. It is the same with an unmonitored wind turbine. A monitored turbine, on the other hand, will do its job until the wind speeds are too high, at which point it will veer or furl so as to "spill" the wind. For example, the blades will tilt or slant back, springs and brakes will come into action, blades will hinge and fold, or electrical systems will shut down mechanisms. There is always the outside possibility that a high wind will blow the whole thing down, however. Choose a turbine that has a proven shut-down mechanism.

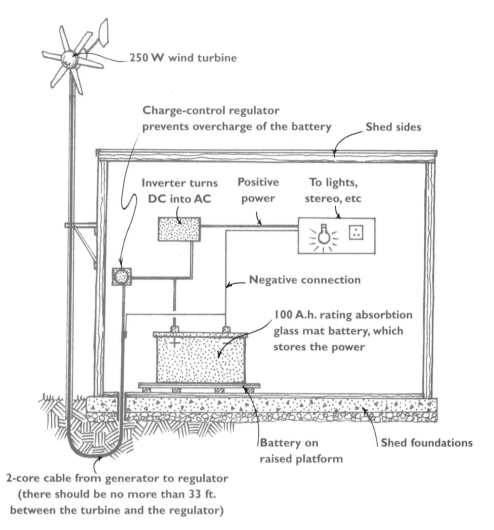

250 W wind turbine

Charge-control regulator prevents overcharge of the battery

Shed sides

Inverter turns DC into AC

Positive power

To lights, stereo, etc

Negative connection

100 A.h. rating absorbtion glass mat battery, which stores the power

Battery on raised platform

Shed foundations

2-core cable from generator to regulator (there should be no more than 33 ft. between the turbine and the regulator)

A small 12 V system

BATTERY STORAGE

If, for example, you want to illuminate your outbuildings using on-grid–type power, you will need 1–2 batteries—with, say, 100 ampere-hour (A.h.) rating—a charge-control device, and a modified sine-wave inverter of around 1000 W.

WHAT WILL THIS SYSTEM POWER?

If all goes well, and the wind blows, a system of this size will power five low-energy bulbs for about 20 hours, a TV, or, in conjunction with the inverter, any number of other small appliances for as long as the battery holds.

COUNTRY HOUSE SCENARIO

In this case, imagine you are a DIY enthusiast, you live on a small property in an isolated spot on the edge of a small, remote community—completely off-grid—and you want to install a 1 kW wind-turbine kit to use alongside your photo-voltaic panels. The goal is to make a good part of your own electricity. While you are aiming to achieve a hybrid standalone system made up of a wind turbine, photovoltaic panels, and bottled gas with a small diesel generator as a backup, central to your plans is the need to keep costs down.

QUESTIONS TO ASK YOURSELF

- Have you researched the wind speeds in your area?

- Have you researched total costs (turbine, tower/mounting brackets, controller, inverter, and batteries)?

- Will you need help erecting the tower?

PROPERTY SIZE AND LOCATION

A bigger turbine would be a better option in terms of power produced. However, while a 1 kW turbine may struggle to keep up with your needs, it wins on cost and ease of erection. For example, with a large turbine you would have all manner of problems such as transport to the site and the need for cranes and a team of helpers; a smaller turbine sidesteps most of these problems.

INSTALLING THE DIY-KIT TURBINE

There is at least one very good company, based in the U.K., which markets a beautiful 1 kW turbine that is specifically designed to be installed by an averagely competent DIY worker. The makers claim that this setup can be installed with nothing more complicated than a range of standard power tools and hand tools. The kit turbine can be mounted by two people in a pinch, but it is easier with a family of four. All the necessary components (see below) can be walked to the site.

Photovoltaic panels and wind turbine

WHAT COMPONENTS WILL I NEED?

The kit manufacturers supply the turbine and the various cables and controls, but you will have to supply a 20 ft. scaffold pole, the concrete slabs that form the foundation base and the guy-wire anchors, and all the various bolts. These component parts can easily be put together from store-bought items.

WHAT WILL A 1 KW TURBINE POWER?

This turbine will power lights, a computer, and a number of low-power appliances. You could power low-voltage lighting from a 12 V system simply by having a few batteries and a string of 12 V halogen lights. Alternatively, you could fit an inverter to power most household appliances. If you want to keep it really basic, you could stay with the photovoltaic panels for the lighting and small appliances and use the turbine for heating water. In this scenario, you spare the expense of batteries and the inverter and simply wire the turbine directly into a low-voltage water-cylinder immersion heater.

CONCLUSION

As ever, just about everything depends upon your unique situation—the precise location of your home, your physical fitness, the availability of help and supplies—but this project should be achievable by any enthusiastic DIY worker.

A country house wind turbine

This project is for you if you live in a small country cottage, completely off-grid, and you want to install a 1000 W, tilt-up/tilt-down wind turbine for lighting and small appliances. You have limited funds and want to go for a low-cost DIY option.

FREQUENTLY ASKED QUESTIONS

- **Can I connect straight into my main electrical circuit?** Yes, but only if you run the power through a grid-tie inverter and install a control charger.

- **What can I power with a 1 kW turbine?** You can store the power in batteries and use it directly for lighting, and/or you can run it through an inverter to run small appliances.

- **Will I need city/county permits?** This depends on where you live, so always check with your local authority.

- **Can I turn the DC power into AC?** Yes, a small DC/AC inverter will do the task.

THE COMPONENTS

A turbine of this type and size usually comes as a kit, complete with the turbine, a set of blades, a tilt-up/tilt-down tower, guy wires, power cables, an inverter, a controller, a bank of batteries, and various fuses and switches. Sometimes, and this will depend upon your chosen option, the tower and the batteries are sold as separate items.

THE TOWER

In simple terms, the higher the tower, the more power the turbine is likely to produce. When it comes to positioning the tower, the ideal is an area of high ground well away from trees and tall buildings. The radius of the circle around your tower needs to be equal to 1.5 times the height of the tower, so a 20 ft. tower will need a circle of ground that measures about 60 ft. across the diameter. Take a 60 ft. length of twine and five pegs and draw circles on the ground until you find what you consider to be a good site.

THE TILT TOWER AND GUY ANCHOR FOUNDATIONS

Depending upon your particular choice of turbine, the guy wire and tower foundations might be anything from metal stakes that you simply hammer into the ground to sheet-metal plates complete with bolts that need to be set into concrete blocks. Most small guy-supported towers need five foundation blocks in all: a large one for the tower, and one each for the four guy anchors that quarter the circle (see page 73). See also how, just prior to the final lift, the turbine end of the tower is eased onto a platform, so that the turbine and blades can be bolted into place.

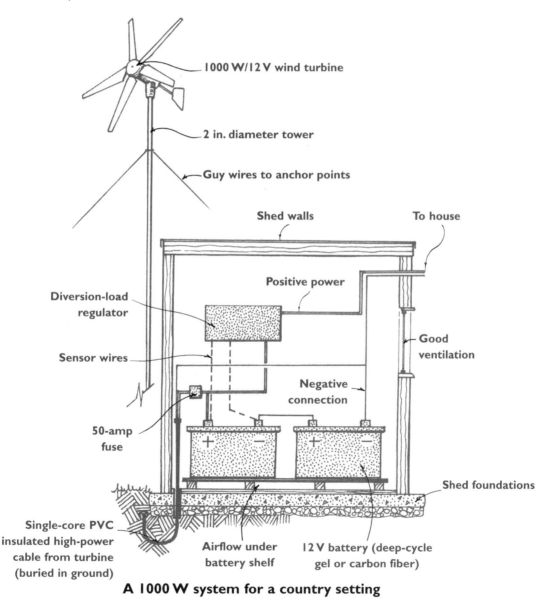

1000 W/12 V wind turbine

2 in. diameter tower

Guy wires to anchor points

Shed walls

To house

Positive power

Diversion-load regulator

Sensor wires

Good ventilation

Negative connection

50-amp fuse

+ −   + −

Shed foundations

Single-core PVC insulated high-power cable from turbine (buried in ground)

Airflow under battery shelf

12 V battery (deep-cycle gel or carbon fiber)

A 1000 W system for a country setting

FARMHOUSE SCENARIO

Here you live on a large property in the country, you are completely off-grid, and you have decided that you want to install a 3 kW wind turbine to make your own electricity. The primary concern here is the physical size of the machine and the impact it will make on your surroundings. Note that the tapered-tower option will involve extra transport cost, more or less double the expense of the tower, will dramatically increase the size and depth of the tower foundations, and will require that the tower be put up by crane.

QUESTIONS TO ASK YOURSELF

- Have you researched the wind speeds in your area?

- Have you researched total costs (turbine, tower/mounting brackets, controller, inverter, and batteries)?

- Have you allowed for additional costs of transport and installation?

- Is the site easy to get to for a truck and a crane?

- Have you allowed for the cost of a ground survey?

PROPERTY SIZE AND LOCATION

Everything about this scenario is big, so it only works if you have plenty of space and are able to hire help. The turbine should be sited as near as possible to the house, but away from anything that is going to get between it and the wind, such as trees, buildings, hills, or electricity towers. The soil structure must be sound, with no loose sand, surface water, shingle, or clay. You will need to have the site surveyed by a qualified ground and engineering surveyor. Ideally, the circle of ground around the tower needs to have a radius equal to 1.5 times the height of the tower.

THE FOUNDATION BLOCK

The taller the tower and the bigger the turbine, the bigger the foundation needed. One maker recommends a block 5 ft. deep and 4 ft. in diameter for a 3 kW turbine.

Farmhouse with solar panel and wind turbine

THE BATTERIES, CONTROLLER, AND ANEMOSCOPE (WIND VANE)

The batteries need to be housed in a dry, cool building. Once they are in place, they should be wired up in series, with the plus-terminal of the first battery linked to the minus-terminal of the second battery, and so on. The manufacturers will generally supply battery-setup details to suit your specific machine. The controller will continuously check the state of the batteries; if the voltage is too high or too low, the controller will shut down the system. The anemoscope will check the wind speed, face the turbine into the wind, and—if the wind speeds are too strong—furl the turbine so that it shuts down.

Under no circumstances should the turbine be run without a load (this means being connected to appliances, or at least to a bank of batteries)—to do so could be dangerous to you and damaging to the system.

HYBRID SYSTEM

In this situation, you will need to create a hybrid system to cover all eventualities. The photovoltaic panels will produce energy for the greater part of the summer, a diesel generator will take over in emergencies, perhaps a wood-fired stove or a bottled-gas system will be fitted as a backup, and so on.

CONCLUSION

Wind turbines of 3 kW and above are serious pieces of equipment, and the setup needs to be done by a specialist. Certainly, the average person could perhaps organize the building of the foundations and the battery house, purchase the batteries, liaise with the crane company, and so on, but the actual setup of the controller, inverter, and displays is best left to experts.

A farmhouse wind turbine

This project is for someone who lives in a good-sized, completely off-grid house in the country and who wants to install a 3 kW turbine on a self-supporting tapered tower as part of a hybrid system. While you have enough money for every option, you want to be involved in some part of the work.

THE HYBRID SYSTEM

With a hybrid system of this size and character, if the wind stops the solar photovoltaics take over, if both the wind and sun fail the diesel generator takes over, and so on. A diesel generator is anything but a green, renewable option, and the same goes for the bottled gas, but if you want an off-grid system that answers every eventuality you will have to accept this.

THE TURBINE

There are many turbines to choose from. The average low-priced 3 kW turbine comes with an anemoscope, a controller, an inverter, and all manner of controls and cutouts. The rotor diameter from blade tip to tip is about 16½ ft. There is a choice between low-speed and high-speed blades. The head automatically adjusts according to wind readings, it has a rated wind speed of 33 ft. per second, a voltage of 280, a power rating of 3000 W, a start-up speed of 10 ft. per second, and the blades furl at a wind speed of 40 ft. per second, which is about 27 m.p.h.—you get a lot for your money.

THE TAPERED TOWER

While a tapered, self-supporting tower is undoubtedly a very tidy option, you must be aware of the implications. The initial cost of such a tower is more than double that of a guy-wire supported tower of the same height. Transport costs are higher because a tapered tower comes as a single piece. The setup costs are much higher because of the need for deeper and wider foundation blocks, and of course, a crane is needed to set the tower in place.

THE CONTROLLER

The controller monitors the voltage of the battery bank and either sends power from the turbine into the batteries to recharge them or, if the batteries are fully charged, dumps the power from the turbine into a secondary load. The controller will read the speed of the wind and the state of the batteries, and in response to the wind speed being too strong and/or the batteries too full, it will furl the turbine and/or shut down the system. When the voltage is over or under a preset figure, the turbine will shut down.

THE BATTERY BANK AND INVERTER

The bank of batteries allows the electrical energy to be stored. In simple terms, the electricity is stored at low pressure—low voltage—in DC form. When you switch on an appliance, the inverter converts the batteries' DC low voltage to "regular" AC higher voltage. Ideally, the batteries should be stored in a cool, dry, well-ventilated shed. The quantity, type, and size of battery will depend upon the size of the turbine and your needs. There are three common setups: a 3 kW turbine using eight batteries with a total rated capacity of 1080 A.h.; a 3 kW turbine using 20 batteries with a total rated capacity of 2000 A.h.; and a 2 kW turbine using 10 batteries with a total rated capacity of 1500 A.h. In all instances, the batteries are connected in series (see page 69).

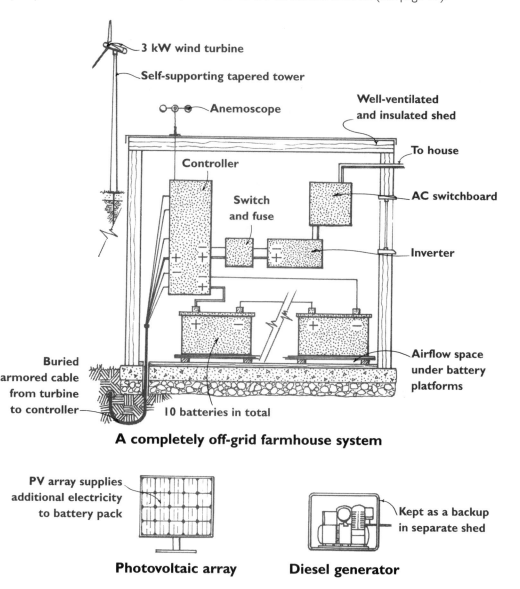

3 kW wind turbine
Self-supporting tapered tower
Anemoscope
Well-ventilated and insulated shed
To house
Controller
AC switchboard
Switch and fuse
Inverter
Buried armored cable from turbine to controller
10 batteries in total
Airflow space under battery platforms

A completely off-grid farmhouse system

PV array supplies additional electricity to battery pack

Kept as a backup in separate shed

Photovoltaic array　　　**Diesel generator**

INSTALLING A GUYED TOWER

When your wind turbine has arrived, you will need a team of people to help you raise the tower. Gather your helpers around and explain that the whole operation is potentially extremely dangerous. Make it clear that there will be just one person in charge, and that the rest of the team must follow the leader's directions at all times. Give each member of the team a single specific task.

MARKING OUT THE GROUND

Take your tape measure, rope, and pegs and mark out on the ground the position of the tower base and the four guy-wire anchors. If you have got it right (using the analogy of a clock face), the tower base will be at the center, the tower will be perfectly aligned between the back 12 o'clock high anchor, the center of the clock and the 6 o'clock front "pulling" anchor, and the two side anchors will be at 3 o'clock and 9 o'clock.

THE FOUNDATIONS

The hole for the tower foundation measures 31 in. square and 62 in. deep, while the anchor holes are each 31 in. square and 3½ ft. deep. If your ground is seriously swampy or sandy, pick another spot, and/or call in the services of a ground engineer. Having worked out how you are going to position the various metal fixings, mix the concrete to the approved standard—1 part cement, 2 parts sand, 4 parts aggregate. Then set the metal fixings in place and shovel the concrete into the holes. Make sure when you set the metal baseplate and the anchors in place in the concrete that they are set at the correct recommended level/angle.

TOWER ASSEMBLY

When the concrete has cured, set the first stage of the tower down on the ground so that its foot is at the central foundation block and its head is looking toward the back 12 o'clock guy anchor, and hinge and lock it in place on the foundation plate. Bolt the second stage of the tower to the first, and then carefully angle the whole thing up from the ground and support the top end on a staging platform so that the top end is about 6 ft. or more off the ground. Pass the cables down the center of the tower, bolt the turbine head in place, fit the various controls as directed, and wire up the cables. Fit the dome/covers. Bolt the four guy wires in place to the tower, and adjust the back wire at 12 o'clock and the side wires at 3 o'clock and 9 o'clock so that they are set slightly slack.

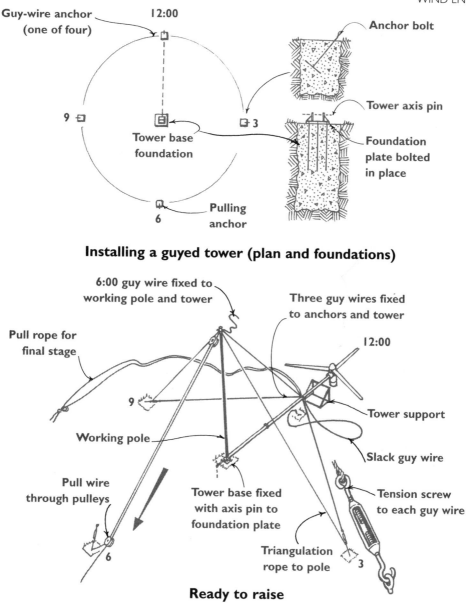

Guy-wire anchor (one of four)

12:00

Anchor bolt

Tower axis pin

9

3

Tower base foundation

Foundation plate bolted in place

6

Pulling anchor

Installing a guyed tower (plan and foundations)

6:00 guy wire fixed to working pole and tower

Three guy wires fixed to anchors and tower

12:00

Pull rope for final stage

9

Tower support

Working pole

Slack guy wire

Pull wire through pulleys

Tower base fixed with axis pin to foundation plate

Tension screw to each guy wire

6

Triangulation rope to pole

3

Ready to raise

RAISING THE TOWER

Slip-fit the single working rope to the front lug. Slide and fix the working pole/gin pole in place and secure it with the two side "triangulation" ropes. Fix the last "pulling" guy wire—one end to its appropriate tower lug and the other to the pole. When all wires and ropes are in place, link the longest "pull" wire, via the two pulley blocks, to your winch or tractor and you are ready to go.

Warning: Everybody must stand far away from the tower in case it falls.

Start up the winch or tractor and slowly winch up or pull away. When the tower is in position, secure the fourth guy wire, make necessary adjustments, and remove all ropes and pulleys.

HINTS AND TIPS

As with all the other off-grid, renewable energy options, the whole area of wind turbines is developing at an astonishing rate—so much so that we will very soon be able to choose anything from a giant turbine big enough to power a village to an array of microturbines that are so small and cheap that we could fit them in clusters on balconies and rooftops.

THINKING IT THROUGH

- **DO YOUR RESEARCH**

  There are so many possibilities that the only way forward is to look at the two primary factors—your needs and your setup (wind speeds, the shape of the land, and so on)—and then make an informed choice.

- **THE RIGHT LOCATION**

  You can have a wind turbine in an urban setting—and this is becoming easier as planning codes are changing and smaller wind turbines are being developed—but the ideal location at this moment still has to be an area of land in excess of half an acre, preferably a piece of raised ground that is completely open on the windward side. As regards city/county permits, you are best advised to make contact with the authorities in your area.

- **ASSESSING THE WINDINESS OF YOUR AREA**

  Consult a wind resource map for your area (usually available on the internet). Keep a daily record, grading the winds as "slight breeze," "stiff breeze," "stiff wind," "very windy," "storm force," and so on, then draw a graph based on your observations. You could also put up a slender mast topped with an anemometer and take readings.

- **MICROTURBINES**

  One of the most exciting developments is in the area of microturbines—ones that are not much bigger than the palm of your hand and can be used en masse. Soon we may be able to have an array of little plastic turbines each making a little electricity, rather than single huge turbines making a lot.

THINKING IT THROUGH (CONTINUED)

- ### BATTERIES
  Research suggests that deep-cycle batteries are the best option—they are designed to give several hundred complete charge/discharge cycles before they need to be replaced.

- ### MONITORING SYSTEMS
  Monitoring systems control the speed of the rotor. When the system judges that the wind speed is too fast, it will cut in and variously swing the turbine around so that so that the blades are out of the wind, and/or it will tilt the blades. Without some sort of monitoring system the turbine would, sooner or later, run catastrophically out of control.

- ### ARE WIND TURBINES NOISY?
  There is no denying that wind turbines make noise, but the "swish-swish" might be music to your ears. The only real solution is to search out an up-and-running turbine and experience the noise first hand.

- ### BLADE SIZE
  If you think of the blades as being wind collectors or wind harvesters, then it follows that the bigger the blades, the bigger the input and the bigger the output. Get the biggest and best that you can afford.

- ### MAINTENANCE
  Just like any other piece of dynamic machinery, such as a tractor or a lawn mower, a wind turbine will need regular, routine maintenance. Bolts will need to be tightened, bits and pieces will have to be painted, some moving parts will need to be greased, the tower and guy wires will have to be checked, and so on.

- ### HOW LONG WILL A TURBINE LAST?
  Research suggests that a good-quality, well-maintained turbine will last anywhere from 20 to 50 years. For example, some old turbines built in the 1950s are still going strong.

Wood-fuel energy

BASICS

Wood basics

From one country to another, there are big differences in terminology. One person's "boiler" is another's "furnace." Read the following carefully and make sure you choose the correct piece of equipment for your needs.

WOOD-BURNING STOVE

This is a standalone stove or heater dedicated to providing space heating. Most wood-burning stoves tend to be decorative, drawing inspiration from traditional black cast-iron stoves. A good modern stove—fired by logs or pellets—is easy to light and control, burns for half a day or more at one filling, keeps going overnight, has a window so that you can view the fire, and has secondary controls that allow you to burn off the tars.

WOOD-BURNING STOVE WITH A WATER HEATER

A stove with the addition of a bolt-on water heater or boiler and perhaps even a soapstone heat-storage slab looks and feels the same as the basic stove—the only difference is that the water heater supplies hot water. If you have two identical stoves, one with a water heater and one without, then the one with a water heater will be less efficient as a room heater, simply because a good part of its output will go into heating the water. You need to take this into account if you want a single wood-burning stove to heat both the room and the water.

COOKING RANGES

Wood-burning cooking ranges come in all shapes and sizes—some very basic, no more than a room heater with a little hotplate and a small oven, and others very fancy with covered hotplates, two or more ovens, pellet feeds, and all kinds of dials and controls.

FURNACE

A utility furnace or boiler is specifically designed to provide heat in the form of hot air and/or hot water. For the most part, they are large, functional, boxlike forms that are designed to be located in a dedicated room, shed, or outhouse.

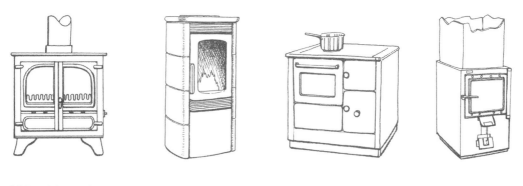

Wood-burning stove   **Auto-pellet stove**   **Wood-burning cooking range**   **Furnace**

WHICH WOOD TO BURN?

The following are the best woods for burning:

- **Ash:** Heavy, difficult to saw and split, long burn, little smoke, high heating value.

- **Beech:** Medium weight, easy to saw and split, medium burn, little smoke, lots of heat.

- **Birch:** Good choice when dry, medium-weight, easy to saw and split, medium burn, lots of heat.

- **Cedar:** Good choice when seasoned, easy to saw and split, smells good, noisy burn, fair heat.

- **Cherry:** Heavy, easy to saw and split, slow burn.

- **Maple:** Easy to saw and split, long, easy burn, good heat.

- **Oak:** Heavy, easy to split, long burn, fierce heat.

- **Walnut:** Heavy, difficult to saw but easy to split, smells good, long burn, good heat.

Other woods that can be used, but are not as good, are alder, aspen, basswood, elm, hemlock, pine, poplar, and willow.

What will it power?

You might think that wood is wood: You burn it, you get direct radiated heat, and that is it. On the contrary, using wood is a surprisingly flexible option. You can't use it to light your home—not readily anyway—but it is fine for just about everything else.

USES FOR WOOD

- Logs can be used to fuel open fires, a simple room heater, or a wood-burning stove.

- You can use wood to heat a stove that is fitted with a water heater or boiler. With a simple gravity-feed system, the hot water rises from the boiler to a storage tank.

- You can use wood in a furnace for heating radiators.

- You can use wood to heat a cooking range, with or without a water heater or boiler, for heating water and radiators and to give background heat.

- You can also use wood pellets (see page 83) to fuel most of the above systems.

ADVANTAGES

- A simple wood-burning stove can be fired up in 5–10 minutes.

- Some people find pleasure in the whole business of sawing, chopping, and fetching the wood—it gives them a sense of well-being.

- A simple wood-burning stove with a water heater can be operated without the need for motors or pumps—a great advantage if you want to live completely off-grid.

- A basic wood-burning system can easily be installed by a clever home DIY enthusiast.

- Wood ashes are good for the garden.

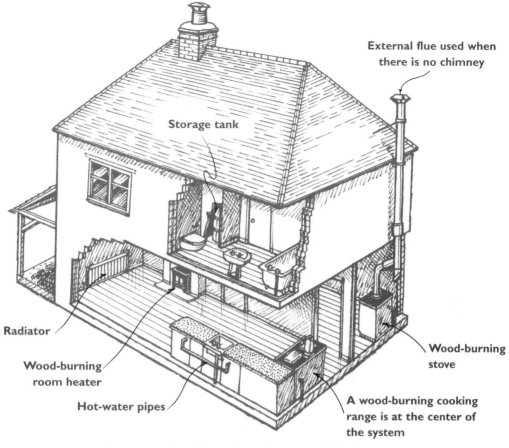

External flue used when there is no chimney

Storage tank

Radiator

Wood-burning room heater

Hot-water pipes

Wood-burning stove

A wood-burning cooking range is at the center of the system

A range of options for using wood fuel

DISADVANTAGES

- Wood—logs or pellets—is anything but instant; its use involves a lot of planning ahead, such as ordering wood and sawing logs so that they are ready for use.

- Traditional log-burning stoves are messy and need a lot of physical input, which is not good if you are old or disabled. They need to be monitored (say every 6–8 hours), even if you are ill.

- It has been estimated that per year the average year-round stove needs about 7 acres of managed woodland to keep it fed.

- A wood-burning system needs storage space for logs.

SCENARIOS

Townhouse scenario

In this scenario, you live in a tiny townhouse, you work from home, and you have decided that you want to heat the house by means of a small wood-burning stove. You prefer modern design to traditional and have decided to fuel the stove with wood pellets rather than logs.

QUESTIONS TO ASK YOURSELF

- Do you want no more than a stove, or could it be, say, a stove with a hotplate?

- Do you want a stove with its own integrated feed hopper or are you happy to direct-feed it with a bucket or hod?

- Do you want a black cast-iron stove or can it be glazed and colored?

- Are there any dimensional restrictions—height, depth, or width? For example, do you have a low mantel that will limit height?

PROPERTY SIZE AND LOCATION

The size and location of your house will, to a great extent, decide your choice. A small yard will limit the amount of pellets that you can store at any one time; a good-sized yard with side access for a car will mean that you can buy in bulk; a traditional house with planning restrictions might make things difficult; and so on.

THE CHIMNEY

Does your house have a chimney, meaning a flue or a smokestack? It is good if the answer is yes; if the answer is no, you will have to build either an internal flue that runs up through the rooms and through the roof, or an external flue that runs straight out through an external wall and up the side of the house. You can either build it with bricks and blocks, finishing with a traditional chimney, or you can put in one of the relatively new stainless-steel double-skin systems. The latter are quick to build, but they can look out of place in an old house; a traditional brick-and-block chimney might look good, but it is a messy, disruptive option that requires a lot of space.

THE HEARTH

Your new stove has to sit on a hearth, meaning a fireproof slab that is large enough to catch ash and fallout from the fire. It can be made from brick, concrete, stone, glass, or steel—anything as long as it is fireproof and conforms in size to the recommended standards as set by your local safety codes.

FUEL—WOOD PELLETS

Wood pellets are supercompressed sawdust. At about ¼ in. in diameter, pellets make for an incredibly efficient burn. A good stove can run for up to three days at a single filling, with the ash pan only needing to be emptied every three weeks or so. Depending upon where you live, pellets can be purchased in bags or in bulk. The storage area needs to be cool and dry.

CONCLUSION

In the context of a townhouse, everything hinges on your local planning and safety codes. Check with the authorities before you start.

A modern wood-burning stove in a townhouse

Country house scenario

This project is for someone who lives in a small traditional house with a large yard in a small town. You work locally, and you have decided that you want a wood-burning stove to heat water and a few radiators.

QUESTIONS TO ASK YOURSELF

- How many radiators do you want to run? Will some need to be disconnected?

- Are you going to cut your own logs, or are you going to buy them in precut? Do you have plenty of storage space for the logs?

- Are you able to tend the fire halfway through the day?

- Do you have electricity for heating water in the summer, or will the stove be in use all year round?

- Do you have space in your house for a storage tank and a hot-water storage cylinder?

PROPERTY SIZE AND LOCATION

Consider carefully how the wood will be delivered, stored, and transported into the house. For example, a property may have good road access but have difficult access to the back of the house, meaning that on delivery the logs may be dumped in the front yard and will have to be moved to the back for storing in a shed, then later moved again into the house. Plan so that work is kept to a minimum.

THE CHIMNEY

Assuming the house has a chimney, check out its size and condition. This is important because the gases from burning wood produce tar that will in time accumulate on the inside of the chimney. This will stain your walls, it will run down the inside of the chimney as a sticky ooze, and, worst of all, it might catch fire. You need to know if your chimney has been lined. Take a flashlight and look up the chimney from the hearth and down the chimney from the roof. Can you see a metal liner or a cylindrical liner, or do you see just a rough rendered surface? If you are starting out with a basic open fireplace, then the chances are that the chimney has not been lined (see page 88 for step-by-step instructions on how to do this).

A cottage-style stove in a country house

THE HEARTH

Your new stove has to sit on a hearth. The easiest option if you are starting out with an open fire is to extend the existing hearth so that it accommodates your new stove.

THE WATER SYSTEM

A simple gravity-feed water system requires that you have a cold-water tank in the attic and a hot-water storage cylinder as near as possible to the stove. The stove heats the water in an integrated water heater, the hot water rises into an enclosed coil within the cylinder, and the resultant hot water within the outer part of the cylinder is used to feed the taps and radiators. The whole system can be run without the need for electric pumps (see page 91).

FUEL—LOGS

Of course, the cheaper option is to saw your own logs, but this option is expensive in both time and energy. The alternative is to buy them precut. Either way, you must make sure that you use the best-quality wood and that the logs are sized so that they fit your chosen stove (see page 79).

CONCLUSION

You will keep the fuel costs to a minimum if you do the log-cutting yourself, but you will need a truck or a car and trailer, a good hand or power saw, a large felling axe, lots of space, and lots of energy. Make sure the system you choose fits with what you have to work with.

Farmhouse scenario

Here you live in a completely off-grid house, such as a small farmhouse, that has good fully lined chimneys but no public water or electricity utilities, and you have decided that you want a wood-burning cooking range to do everything—cooking, heating the domestic water, and running a few radiators.

QUESTIONS TO ASK YOURSELF

- How many radiators do you want to run?

- Are you able to tend the fire halfway through the day?

- Do you do lots of cooking and need several ovens?

- Do you have space in your house for a storage tank and a hot-water storage cylinder?

- Do you have plenty of storage space for logs?

PROPERTY SIZE AND LOCATION

This scenario needs space. With the ideal layout being a header tank in the attic, a hot-water storage cylinder directly below the header tank, and a wood-burning range just below the storage cylinder, you need to look at your house and see if it is going to work. With the knowledge that the system works by gravity and water pressure, you can nudge things around slightly—for example, the storage cylinder can be below the header tank and backed onto the cooking range—but you can't have long pipes running between the range and the storage tank.

THE COOKING RANGE

The wonderful thing is that wood-burning range technology is so well-established that the best course of action is to seek out an older neighbor and see what they advise. In general, you will need to get up early to open the vents on the range and feed in a few logs. Half-an-hour later, when the range is up to heat, you can boil a kettle of water and cook a traditional breakfast if you wish. After breakfast, you can put a casserole-type meal in the oven and leave it to cook—a good range will have controls that will allow you to regulate the temperature easily. At the end of the day, you will need to feed in logs and close down the vents before going to bed.

THE OPEN-VENTED HOT-WATER SYSTEM

With a traditional open-vented hot-water system, there are five primary elements: a large cold-water header tank in the attic, an indirect hot-water storage cylinder at some point below the header tank, the stove, cooking range, or furnace with a water heater, the taps over the sink and bath, and a small number of radiators. The water flow within the system is driven by water pressure and gravity—the weight of water in the header tank is enough to push water through to the taps and radiators.

A wood-burning stove for cooking, heating, and hot water

ADVANTAGES AND DISADVANTAGES

● **ADVANTAGES**
  - This system is easy to install and maintain.
  - It does not rely on electric pumps.
  - There is less to go wrong.

● **DISADVANTAGES**
  - It takes up a lot of space.
  - The tank in the attic might freeze.
  - Large pipes look a little functional and unsightly.
  - The amount of hot water is restricted by the size of the storage tank.
  - In the context of going without electric pumps, the system will only work within tight confines.

SYSTEMS

Installing a wood-burning stove

If you live in small dwelling with a well-built existing chimney, you may decide to install a wood-burning stove to heat the room. You will need to install a flexible liner, build a hearth, and set the stove in place.

INSTALLING A FLEXIBLE CHIMNEY LINER

This is a task best done by three or more people. Before you begin, you must open up the fireplace so that you can look up the chimney and check that it is in good condition, have the chimney swept, arrange a suitable ladder, and, if necessary, organize things so that you can tether yourself to the chimney stack.

Take a length of rope up to the roof, tie a weight to one end, and lower it down the chimney. A helper calls up when they have caught hold of the weighted end of the rope. Tie your end of the rope around the chimney stack and go down for the flexible liner. Together with your outside helpers, pull and push the liner up onto the roof. Having made sure it is the correct way up (painted arrows point up), tie what will be the bottom end of the liner to the rope. Call down to your helpers and tell them to pull on the rope when you give them the go-ahead. Gently feed the liner down the chimney while your helpers gently pull on the rope. After a lot of shouting, pushing, and pulling, when the liner is in place, clamp the top of the liner to the chimney and fit a cowl that best suits the specific design of your chimney. Finally, fill the space between the liner and the top chimney surround with mortar.

BUILDING A HEARTH

The hearth can be made from concrete, stone, brick, steel, glass, or ceramic tiles—anything as long as it is fireproof. If you have a wooden floor, most building and safety codes require that the hearth be at least 9 in. thick, or 5 in. thick if you have a space between the floor and hearth, or about ½ in. thick if the base of the stove is guaranteed not to rise above 212°F. If the floor is solid concrete, you can sit the stove directly on the floor.

INSTALLING THE STOVE

When fitting a stove to an existing chimney, you can either use a back-vented stove (with the flue running straight out of the back of the stove and into the bricked-up chimney) or a top-vented stove (with the flue running straight up the chimney), depending on your fireplace.

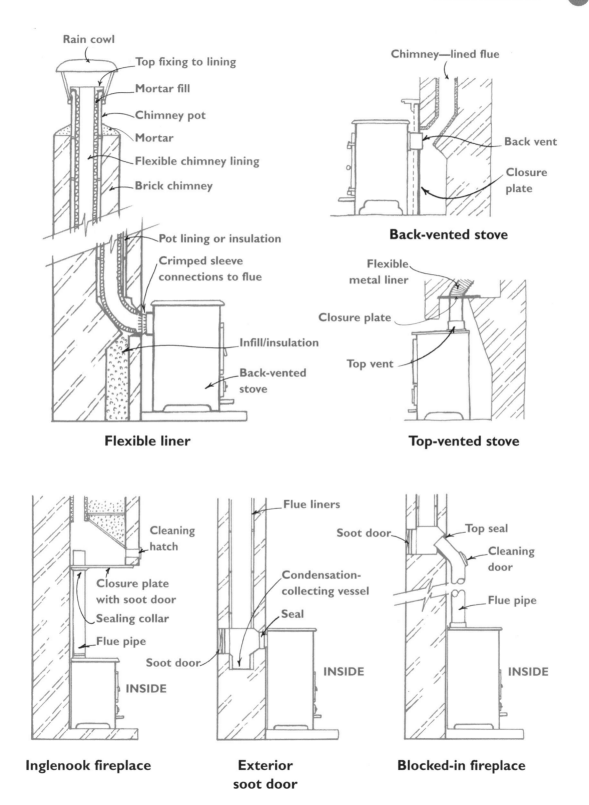

Flexible liner

- Rain cowl
- Top fixing to lining
- Mortar fill
- Chimney pot
- Mortar
- Flexible chimney lining
- Brick chimney
- Pot lining or insulation
- Crimped sleeve connections to flue
- Infill/insulation
- Back-vented stove

Back-vented stove

- Chimney—lined flue
- Back vent
- Closure plate

Top-vented stove

- Flexible metal liner
- Closure plate
- Top vent

Inglenook fireplace

- Cleaning hatch
- Closure plate with soot door
- Sealing collar
- Flue pipe
- INSIDE

Exterior soot door

- Flue liners
- Condensation-collecting vessel
- Seal
- Soot door
- INSIDE

Blocked-in fireplace

- Soot door
- Top seal
- Cleaning door
- Flue pipe
- INSIDE

Installing a wood-burning stove with a water heater

This is ideal for someone who lives in small-town house that does not have a chimney, who works locally, and who has decided that they want a log-burning stove to heat the domestic water for the kitchen sink and the bathroom.

DOUBLE-WALL FLUES

Double-wall flues—they come in kit form—are stainless-steel twin wall tubes that are able to withstand high temperatures and corrosive flue gases. They are expensive, but they are fast and easy to install. They are a very good option if you are looking to build a new flue.

Internal A top-quality internal stainless-steel, double-wall flue is a system that goes from the stove straight up through the floors and ceiling and through the roof. The downside is that a weatherproof seal needs to be fitted at the point where the flue pipe goes through the roof covering—the slates or tiles. That said, you don't need to knock a hole through the wall, it always remains warm with the effect that it achieves optimum draw, and an internal system is less expensive than an external one.

External In some situations—say, if the rooms are very small—then it is best to install a stainless-steel, double-wall flue system on an external wall. The system goes from the stove, through a hole in the wall, and then via a support bracket up the outside wall of the house, up, over, and around the various overhangs—all the boxes and gutters that make up the edge of your roof—and then on up to just past the roof's ridge line.

CLAY-LINED CHIMNEYS

Clay liners, in the form of lightweight terra-cotta pipes or tubes will give you a high-quality, permanent, insulated chimney. However, municipal codes in some areas may prohibit installation of new clay chimney liners and require all new liners to be stainless steel. Check with your local authorities.

THE HOT-WATER SYSTEM

A small, basic stove complete with a water heater will warm the room and give you hot water for the kitchen and bathroom. The illustration on page 81 shows a basic layout for a hot-water system for a small house. Note that a single small radiator is installed to prevent the system from overheating. The most efficient layout is to have the water-storage cylinder as close as possible (vertically) to the stove—for example, to have the stove on the ground floor and the cylinder directly on the floor above. Cold water runs into the bottom of the boiler, then the water heats up and rises the short distance to the cylinder, this in turn allows more cold water to enter the boiler, and so on.

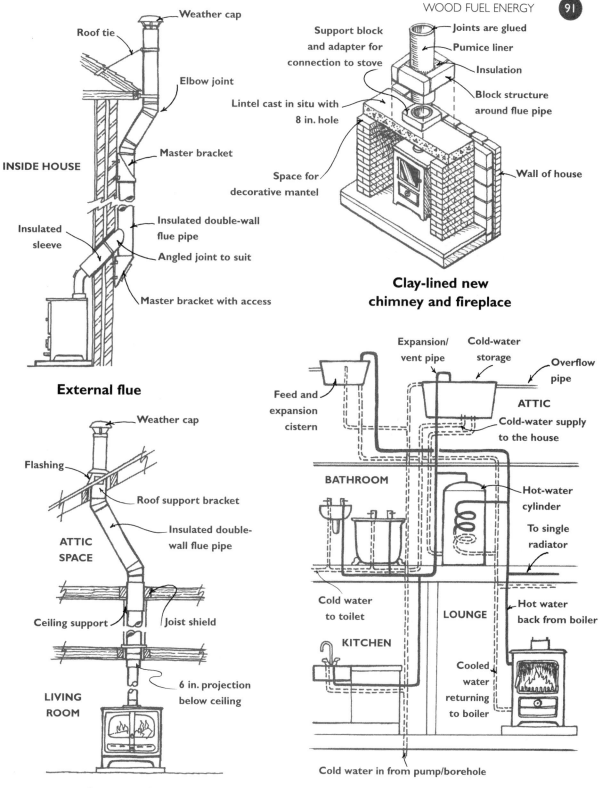

External flue

Clay-lined new chimney and fireplace

Internal flue

Hot-water system

Installing a wood-burning cooking range for heating and hot water

You may live in a completely off-grid house—a small farmhouse or cottage—that has a well-lined chimney in good condition, and you may work in and around the home and have decided that you want a wood-burning range to do the cooking, heat the domestic water, and run a few upstairs radiators.

THE COOKING RANGE

The important thing here is to choose a range that is specifically designed to run on wood, because you can't easily convert a gas or coal range to run on wood. Depending upon the type of wood, the size of the combustion chamber, and the settings, most ranges will need to be fed with logs every hour or so. You will need to be on hand so that you can adjust the controls—meaning open and close the flues—so that you can put the stove in either cooking or heating mode.

THE HEARTH

As with all stoves, the hearth needs to be solid, level, fireproof, and next to a chimney. If the hearth is set between the pillars of the chimney breast, you must make sure that there is a small amount of space all round the range for it to expand and contract. There must be space at the back and sides, and the tiles must be fitted so that they run behind the top plate. In other words, the wall tiles must not overlap the top of the stove.

THE CHIMNEY

As with other stoves, the chimney must be at least 12 ft. high, insulated, and generally in good condition (see page 84).

FLUE CONNECTION AND LAYOUT OPTIONS

With most stove and cooking ranges, the design will allow you to change the position of the flue outlet to suit the shape and structure of the chimney. For example, a rear flue outlet will allow you to set the stove against an existing masonry chimney, while a top flue outlet will allow you to set it in a position with the flue running straight up. Most manufacturers provide technical sheets that specify precisely how the stove needs to be installed. It is vital that you follow the manufacturer's installation directions.

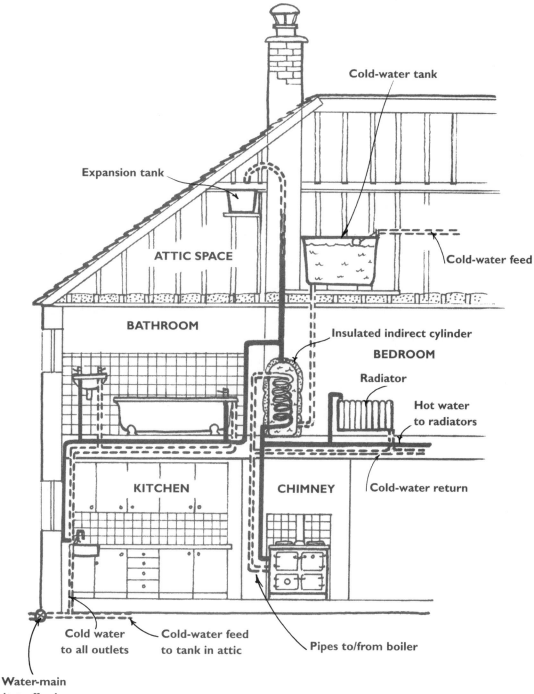

Cold-water tank

Expansion tank

ATTIC SPACE

Cold-water feed

BATHROOM

Insulated indirect cylinder

BEDROOM

Radiator

Hot water
to radiators

KITCHEN

CHIMNEY

Cold-water return

Cold water
to all outlets

Cold-water feed
to tank in attic

Pipes to/from boiler

Water-main
shut-off valve

Wood-burning cooking range and central heating

HINTS AND TIPS

Putting in a new chimney, choosing the right stove for the job, cooking on a wood-burning range, and generally sourcing, choosing, sawing, chopping, and storing wood can be challenging pleasures—but only if you get them right.

THINKING IT THROUGH

- **SAFETY FIRST**

  Install heat alarms and have a dry-chemical fire extinguisher within reach of the stove. Never use a vacuum cleaner to remove the ash from around the fire door—the hot ash could cause the cleaner to burst into flames! Tell children about the dangers of the fire and make sure that, as much as possible, the arrangement of the furniture directs children away from accidental contact with the stove.

- **GOOD WOOD**

  It is best to cut the wood in spring, dry it in summer, and burn it in winter. This is because all wood burns better when "seasoned," meaning thoroughly dried out.

- **A GOOD CHIMNEY**

  If you get the chimney right, the stove will be easy to light, the fire will burn brightly, the smoke will be taken away, and you will not have a smoky backdraft when you open the door to feed in logs.

- **MAXIMUM BURN**

  A poor burn results in lots of smoke, which in turn results in a dirty chimney and air pollution. You should adjust the air intake so that the fire is bright and the smoke barely visible.

- **KEEPING THE FIRE GOING OVERNIGHT**

  Last thing at night, rake the embers toward the air inlet, fill the back of the firebox with logs, and close the air intake down to the minimum. In the morning, put in the logs, open the controls for about 20 minutes so that the fire flares up and burns off the tars, and then set the controls for the day.

THINKING IT THROUGH (CONTINUED)

- **TAR OOZE**

  If you have sticky tar oozing down from one of the joints on the outside of the flue pipe, then you have two problems—you are shutting off the airflow for overly long periods so that the tars are condensing inside the chimney, and you have the male into female joints in the flue the wrong way around. The pipes should be arranged so that liquids are always directed from one pipe down into the next. In this way, tars will run into the stove to be burned off.

- **INTEGRATED WATER HEATERS**

  A water heater will reduce the efficiency of a room-heating stove. If you want a stove to heat both the room and water, it pays to get the largest size you can afford.

- **HEAT-SINK RADIATOR**

  It is wise when installing a water-heating stove—one designed to heat the room and domestic water—to use a single radiator. If the fire gets too fierce and the water gets too hot, the radiator will dissipate some of the heat.

- **GRAVITY AND CONVECTION SYSTEMS**

  These are undoubtedly the best option for an off-grid house because they don't need a pump or electricity, but you must install large-diameter flow and return pipes, and have straight pipe runs, meaning slow curves and 45-degree elbows, rather than 90-degree bends and loops.

- **SMOKY STOVES**

  If the stove suddenly starts to smoke, it usually means that the chimney needs sweeping and/or the seals around the door/flue are broken.

- **ASH**

  Wood fires produce surprisingly small amounts of ash. Remove the ash about every 7–10 days, when it is not raining or windy.

- **FIRE WILL NOT SHUT DOWN**

  A fiercely blazing fire suggests that the seals around the door have broken, or the stove, door, or glass is cracked, or the air control/damper is stuck or blocked.

Bioenergy

BASICS

Bioenergy basics

"Bioenergy" or "biomass energy" is energy made available by combustion of materials derived from biological sources. In other words, bioenergy is energy generated from renewable biological resources.

WHAT IS BIOMASS?

Biomass is plant material of recent biological origin—logs, straw, chipped-up trees, charcoal—but the term also includes all the waste produced by our society, such as industrial sludge left over after food and drink manufacturing, human and animal manures, and household waste. Biomass can be converted into solid, liquid, or gaseous fuel. The character of the base material will decide how it is used as bioenergy. For example, chipped, crushed, and pelleted trees pose a very different energy conversion problem from liquid cow manure.

FOOD-PRODUCTION WASTE

Used vegetable oil can be filtered, chemically treated, and turned into biodiesel relatively easily.

BIODIESEL

Biodiesel will work in just about any conventional diesel engine. At a relatively low start-up cost, you can obtain a store-bought, domestic-sized biodiesel production plant. If you live in a city, you may be able to take old vegetable oil away from restaurants and fast-food outlets and quickly make it into a usable fuel. Note that we do not advocate making engine fuel from food products such as corn, sugar cane, or sugar beet in a world where there are food shortages.

MANURE

In developing communities, cow and goat dung has traditionally been dried and burned for heating and cooking. With industrial-sized cattle and dairy units, manure can be used to produce heat and electrical energy. This only really works on a large scale, however.

WOODCHIPS

Growing woodland crops, such as willows, that can be turned into woodchips is a good option because they have a higher and faster yield than mature tree forests, and they can be grown on poor, low-lying ground that would be otherwise useless.

FREQUENTLY ASKED QUESTIONS

● **Can I use cow manure to produce natural gas?** While in theory you can, this only really works on a dairy-farm scale. In this case, it would be relatively easy to install a methane digester to extract gas from cow manure. Perhaps you could make it work by joining forces with a group of neighbors.

● **Can I use kitchen and toilet waste to create energy?** Yes, you can install a composting toilet to turn all your kitchen and toilet waste into garden compost.

● **Is wood alcohol an option?** It is easy enough to make wood alcohol on a small scale in a laboratory, but it is not really viable as a fuel. If you have a large quantity of wood in chip form, it is much easier to burn it in a stove.

● **What is biogas?** Biogas is created by digesting food or animal waste in the absence of oxygen, a process that only really works on an industrial scale, but you could possibly make it work if you live in a small town (see pages 162–71).

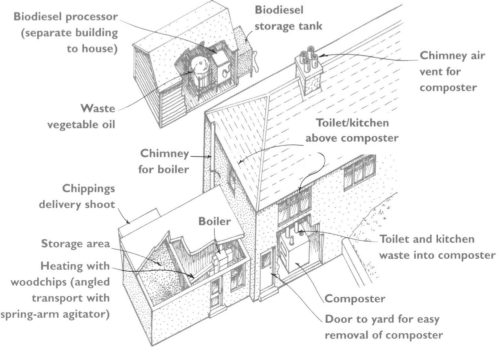

Bioenergy options

What will it power?

Although bioenergy can, in its various forms, be used to power just about anything, a big problem in the context of a domestic off-grid system is scale. For example, cow manure can be turned into gas, but you need a lot of it and a large setup. Sawdust can be burned, but it is difficult to transport, and a furnace system complete with a machine to make pellets, a hopper, and an automatic feed from the hopper to the furnace will be an expensive setup. The best low-cost options are usually a composting toilet, a woodchip-burning furnace, and a plant to turn waste cooking oil into biodiesel to run a generator to produce electricity.

BIOENERGY USES

- You can use logs to fuel open fires.

- You can use logs to fuel a sophisticated cooking range with integrated water heater—this would heat your rooms, give you hot water, and allow for cooking (see pages 90–91).

- You can use woodchips to fuel a large water heater and a central-heating system—this would heat radiators and hot water.

- You can use sawdust, grass, straw, or paper in pellet form to fuel a central-heating system—the setup would require that you have a machine to make pellets, but it would be automatic.

- You could turn vegetable-oil waste into biodiesel—this would fuel your vehicle and a generator and could be traded to friends and family.

- You could have a composting toilet—the compost would enrich your land.

- You could join forces with close neighbors and run liquid animal manure through a methane digester—a town-sized setup would give a small community power, heat, and light.

ADVANTAGES AND DISADVANTAGES

● **ADVANTAGES**

- A simple wood-burning stove can be fired up in 5–10 minutes and is a great choice if you live in an area where there is plenty of wood in the form of logs. The setup costs are relatively low.
- Some people find an enormous amount of pleasure in and derive a sense of well-being from the whole business of sawing, chopping, and fetching the wood.
- You can be pretty certain, if you live in a forested area, that supplies of wood will stay constant.
- Biodiesel is a good option if there is a ready supply of waste vegetable oil in your local area.
- Biodiesel can easily be used to fuel a vehicle.
- Biodiesel can be used in a generator for electricity.
- Some countries give grants to help with small-scale biodiesel production.
- Burning waste wood, grass, or straw in pelleted form is a good option if you are looking for a near-automatic system.

● **DISADVANTAGES**

- Wood (logs or pellets) is anything but instant—its use involves a lot of planning ahead.
- Traditional log-burning stoves are messy and need a lot of physical input.
- It has been estimated that per year the average year-round stove needs about 7 acres of managed woodland to keep it fed.
- A traditional log-burning system will need to be constantly watched and managed, around every 6–8 hours, even when you are ill.
- A log-burning system needs lots of storage space—sheds, lean-tos, buckets, and baskets.
- The setup for a wood, grass, or straw pellet-burning system is expensive, and you have the problems of transport and storage.
- Biodiesel requires special storage and, in some areas, a special license. Some countries also levy a tax on its production.
- There is always the chance, with something like woodchips, waste oil, or sawdust, that supplies will dry up.

COMPOSTING TOILETS

Installing a composting toilet

In this scenario, you live in the country and you have decided that you want to install a composting-toilet system. You want to turn all your kitchen and toilet waste into compost, and you want to cut down on the amount of valuable water that you put down the drain.

QUESTIONS TO ASK YOURSELF

- Do you need city/county permits?

- Are there any restrictive building codes in your area?

- Is your house suitable (meaning the levels and the way the rooms are organized)?

- Have you researched the total cost of the system and its installation?

PROPERTY SIZE AND LOCATION

The size, design, and location of your house will always, to a greater or lesser extent, shape your off-grid, renewable-energy options, even more so if you want to install this type of composting toilet. Your property needs to be: a split-level design on a sloping site; a house or bungalow with a large cellar or basement; or an "upside down" house with the kitchen, living area, and bathroom upstairs. In other words, the layout must be such that your kitchen and toilet waste can drop down into the composting chamber, installed at the lowest part of the house, and there must be easy access to this chamber.

COMPOSTING TOILETS AND SMELLS

The design of this type of composting toilet is such that the downdrafts that whip down the toilet and up the chimney, and down the kitchen waste hatch and up the chimney, are so strong that all the smells are quickly removed. Some models are fitted with fans to increase the flow of air.

Composting
toilet

An "upside down" house with a composting chamber

IS IT SAFE?

Research suggests that, of all the many installations, there has never been a problem with safety. The kitchen and toilet waste breaks down into a fine, crumbly, brown tilth that is completely safe to handle—very much like potting compost.

DOES THE PROCESS NEED ANY WATER?

It needs a little water, but no more than you would expect to get from urine and kitchen scraps—things like peelings, teabags, and mealtime leftovers.

POTENTIAL PROBLEMS

Research suggests that when people have trouble with flies, for example, the problems are caused by sloppy management and maintenance. As with most dynamic systems, you must follow the user guides to the letter.

CONCLUSION

This type of composting toilet is an excellent choice if everything is right—the design of your house, the site, the lay of the land, your lifestyle, and so on. It does the job without the need for complicated electrics, it has very few moving parts, and it has a proven history.

A country cottage composting toilet

This project is for someone who lives in the country and wants to lower their impact on the environment, lower their water usage, and enrich their vegetable garden all by installing a composting toilet. Composting toilets have been around for a long time. It is a wonderfully basic idea: You do your business, you leave it for a year or two so that nature can do its work, and then you use the resultant compost to enrich the land.

HOW DO COMPOSTING TOILETS WORK?

Most composting toilets, depending upon their size, make, and design, have five primary elements: an entrance for the waste—the toilet seat and the kitchen-waste lid; a chimney topped with an exhaust valve; a composting chamber; a sloping bed; and a removal hatch. If you look at the illustration on page 105, you will see that the waste drops down, it gradually heaps into a pile, the excess water drains off to the liquid chamber and/or evaporates, the drying heap molders and gets hot, and as the heap is being microbiologically broken down into compost, it gradually slides down a slippery slope toward the removal hatch.

PLANNING THE LAYOUT

One look at the illustration will show you that just about everything hinges on three factors: The toilet waste needs to be able to drop into a chamber; the heap in the chamber needs to be able to slide down a slope; and at the bottom of the slope there needs to be a door or hatch for compost removal. Depending upon the design, the total vertical drop from the top rim of the toilet pan down to the concrete floor on which the whole system sits is about 7–8 ft.

BLACK WATER

"Black water" is water that we put into the toilet in our urine and the watery part of our feces. The small amount of black water that sometimes oozes from the composting system in the early stages can be run into a separate standalone, underground septic-type tank.

INSTALLATION TIPS

- The moldering heap must be warm, or at least not cold. The composting action only works if it is warm. For this reason, some cold-climate systems have built-in electric heaters to speed up the natural composting action.

- The whole working action depends upon there being plenty of fresh air and a strong downdraft—so that smells are drawn down the toilet pan, down the kitchen hatch, and up the exhaust chimney. The higher the chimney, the better the action. Some designs have electric fans to push and/or pull the air around the system.

- While systems like this one are large enough to allow 2 to 3 years for the cycle, some smaller systems have a range of heaters, fans, paddles, and stirrers to speed up the action.

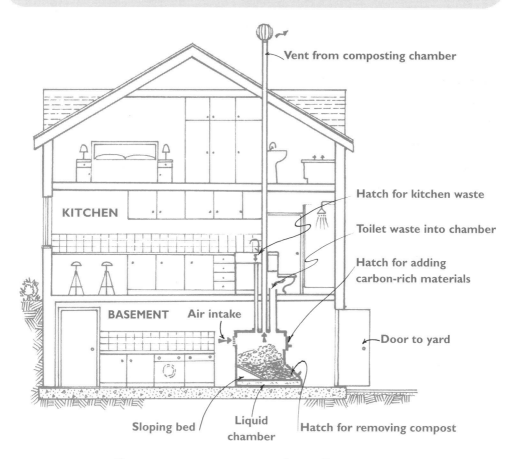

Vent from composting chamber

Hatch for kitchen waste

Toilet waste into chamber

Hatch for adding carbon-rich materials

Door to yard

KITCHEN

BASEMENT   Air intake

Sloping bed   Liquid chamber   Hatch for removing compost

Country cottage composting-toilet system

WOODCHIP SYSTEMS

Installing a woodchip furnace

If you live in an area where there is a plentiful supply of woodchips, you may decide that you want to install a woodchip-fired furnace for central heating.

QUESTIONS TO ASK YOURSELF

- Perhaps the most important question of all is do you live in an area where the supply of woodchips or pellets is guaranteed?

- Are there any restrictive building codes in your area that have to do with storing fuel?

- Is your house suitable (meaning the levels and the way the rooms are organized)?

- Have you researched the total cost of the furnace, woodchip storage, and transport?

PROPERTY SIZE AND LOCATION

It is unlikely that there will be a problem with the shape and size of your house—there are woodchip-fired systems for just about every building configuration—but remember that a constant supply of the woodchips is essential.

A FURNACE IN THE BASEMENT

You could have the furnace in the basement and the fuel stored in the room above or in an adjoining building.

STORING THE WOODCHIPS

A common method is to have the delivery hopper on the side of the house with a chute running down to the basement, with a fuel screw conveyor (rotating helical-screw device for moving fuel along a tube) running from the basement up to the boiler in the room above. Another option is to have the fuel stored in a building that is remote to the house. For example, you could have the storage bunker by the gate, the fuel screw conveyor running through some plywood-box ducting, and the boiler in a small lean-to building at the side of the house; the distance between the storage bunker and the house can be up to 50 ft.

House with woodchip-fired heating system

IS THE SYSTEM WELL-PROVEN?

Yes—some woodchip systems have been around for well over 20 years.

IS THE SYSTEM AUTOMATIC?

If the storage system is big enough and the various automatic controls are in operation, then the system is completely automatic. Some systems are designed to be run remotely—say, you and a computer in one location, and the furnace and fuel supply in another.

CONCLUSION

In just the same way as wind turbines need wind and water turbines need water, there is no getting away from the fact that a woodchip system needs woodchips. The only difference this time is that while you can more or less know that the wind and water are always going to blow and flow, you cannot say the same of woodchips. The problem is that, while not so long ago woodchips were a waste product, they are now becoming a valuable commodity fuel for small-community power stations, the raw material for various building-board products, used in agriculture, turned into biodiesel, and so on. The big question is whether your supply will be guaranteed for the foreseeable future.

A woodchip heating system

This is a good project if you live on a property in an area where there is a guaranteed supply of wood in chip form, and you want to use the chips to fuel a large central-heating system. For some people, the whole notion of wood-fired heating systems sounds messy—lots of logs and too much dust and physical work. For other people, it is the hands-on physicality that makes wood-burning furnaces so attractive. The good news, for those folks who like it clean and automated, is that there are now some very neat domestic systems specifically designed to burn wood in chip form.

WHERE DO WOODCHIPS COME FROM?

Wood producers are left with mountains of waste wood—shavings, chips, and sawdust. This waste is now turned into uniform chips or pellets, which can be transported in bulk to the point of use. You will need to do some research to find a reliable woodchip supplier in your local area.

HOW DO WOODCHIP SYSTEMS WORK?

The systems are beautifully simple. You have a storage silo or hopper for the chips, a furnace for burning, and a fuel screw conveyor (see the illustrations) for moving the chips from their storage to the furnace. The wonderful thing about this system is the fact that there are so many storage-to-furnace options. Storage can be in a room above or below the furnace, in a room next door, across the yard, or in just about any situation where it is possible to run a conveyor from the storage to the furnace.

WILL THIS SYSTEM BURN WOOD IN ALL ITS VARIED FORMS?

Certainly the conveyor is only designed for chips, pellets, and a mix of chips and dust, but there are now systems that are designed specifically to burn logs en masse. You just load up the furnace from top to bottom with logs, and then leave it for the next 20 hours or so.

LAYOUT CONSIDERATIONS

The starting point is deciding where to put the hopper. A huge truck will have to leave the road, pass through your gate, maneuver to your hopper, and make the drop. The next question is whether the truck is going to tip the chips down a chute or whether the chips will eventually need to be at a higher level. If the latter, will the truck have to be fitted with some sort of blowing system. Or will your system have a pit or hopper for first-stop storage and a high-level silo or hopper complete with its own screw-drive channel for primary storage?

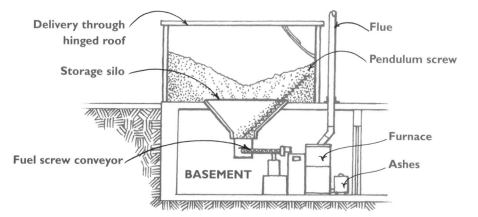

Woodchip heating system: above the furnace

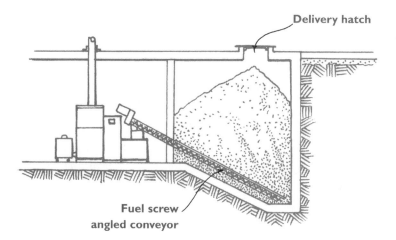

Woodchip heating system: below the furnace

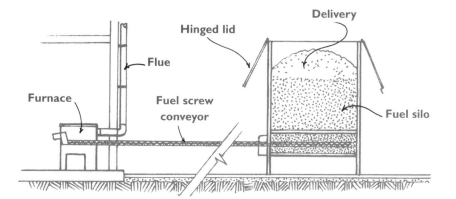

Woodchip heating system: beside the furnace

BIODIESEL

Making your own biodiesel

In this case, you might live on the edge of a small town and decide that you want to make biodiesel (see page 112) for your own use—to fuel your vehicles, for central heating, and as a standby diesel generator.

QUESTIONS TO ASK YOURSELF

- Are there any restrictive building codes in your area that have to do with making and storing liquid fuel?

- Is your property suitable (meaning is there enough space and suitable outbuildings, and is it far enough away from neighbors)?

- Have you researched the total cost of the basic systems, the storage tanks, and all the other items that make-up the system?

- Are you prepared to routinely put some part of your time into the production process?

PROPERTY SIZE AND LOCATION

The big issue with making your own biodiesel is safety and space. You cannot live in a small inner-city house and make biodiesel in your basement—it would almost certainly contravene all kinds of heath and safety codes. Also, once you have made a good quantity of biodiesel, it needs to be stored somewhere.

CAN BIODIESEL BE MADE AT HOME?

Given the right equipment, it is a relatively basic process to make biodiesel—from either straight vegetable oil (SVO) or waste vegetable oil (WVO). It is now possible to obtain very neat, low-cost processors in sizes that range from huge operations to small domestic ones that will fit in a garage.

Biodiesel plant

House with biodiesel plant

BIODIESEL IN DIESEL CAR ENGINES

As to whether biodiesel can be used in your car, much depends on the car, the age of the engine, and the quality of the biodiesel. You might have to install high-capacity filters, but it should work in any diesel engine. Research suggests that the use of biodiesel may, in the initial period of use, release deposits from tanks and pipes which blocks the filters. The best thing to do is have a really good burn with the biodiesel and then change the filters before continuing.

CAN THE BIODIESEL BE STORED IN THE PRODUCTION PLANT?

There is no reason why, if you are making the diesel for your own use, you cannot store the diesel in the plant, but it would be a much better idea to pump the diesel to a holding tank for reasons of safety and convenience.

TAX IMPLICATIONS

You may be required to pay a tax if you make large amounts of biodiesel every year and if you are making it for resale. Always research the codes and regulations in your area.

CONCLUSION

Biodiesel is a good off-grid option if you have a ready supply of oil, if you understand all the transport and storage implications, and if you like the idea of rolling up your sleeves and getting involved in the production.

A biodiesel production system

This system would be ideal for someone who wants to produce biodiesel for their own use and who lives on an edge-of-town property that has plenty of outbuildings with space around them in an area where there is a steady supply of waste vegetable oil. In the 1890s, when Rudolf Diesel invented his engine, he designed it specifically to run on vegetable oil, although things changed in the 1920s when low-cost petrodiesel came into being. This shows that biodiesel is not an inferior fuel, but was and still is the best choice.

GETTING IT ALL RIGHT

The system will only work if everything is right—the supply of waste oil, your interest in recycling, your knowledge of your locale, your need for the fuel, your transport, and, perhaps most importantly, the setup of your property. You will need to get this checked to make sure everything will conform to local health and safety codes. There is also the problem of what to do with the glycerin byproduct.

Arm yourself with a question list, and then make contact with one of the manufacturers. Depending upon where you live, they will answer all your questions. For example, one company sells six different processors ranging in capacity from 40 gallons to 530 gallons (150 to 2000 liters), they can put buyers in contact with oil suppliers, they are prepared to custom-design units to suit your setup, they will dispose of or sell your glycerin byproduct, and they offer one-day courses.

MAKING BIODIESEL

A typical biodiesel production plant—the sort of thing that you can buy as a kit—is made up of three primary elements: a holding tank for the waste vegetable oil, a processor with a heater and mixer, and a delivery tank with a pipe and gun. In addition to the oil, you will need supplies of methanol and lye (sodium hydroxide). If using a kit, you should follow the manufacturer's instructions at all times. From start to finish, the procedure is as follows:

- Pour and filter the waste vegetable oil into the processor.
- Heat the oil up to 120°F.
- Analyze the oil to determine the amounts of free fatty acid and how much lye you will need.
- Mix the lye and methanol together, pour it into the processor, and whisk.
- Allow the mix to separate.
- Remove the layer of glycerin.
- Mix water with the biodiesel, stir, and allow the oil and water to separate.
- Remove the water and transfer the biodiesel to the storage container.
- Allow the biodiesel to dry, and it is ready for use.

HEALTH AND SAFETY

While biodiesel is relatively easy to make, there are health and safety issues that you need to address. For example, methanol is both flammable and toxic, lye is caustic, and at various stages throughout the procedure the mixture will give off toxic fumes. You will need to make sure that your workshop or outbuilding is fireproof and well ventilated, and that you are outfitted with protective clothing, gloves, and eyewear.

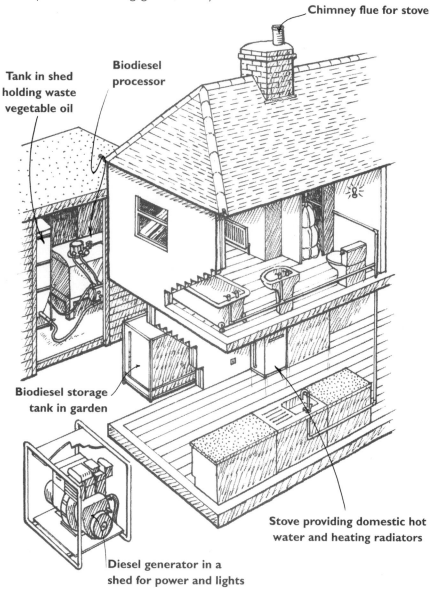

Chimney flue for stove

Biodiesel processor

Tank in shed holding waste vegetable oil

Biodiesel storage tank in garden

Diesel generator in a shed for power and lights

Stove providing domestic hot water and heating radiators

A biodiesel system

HINTS AND TIPS

Systems like a composting toilet, a woodchip-fueled central-heating system, or a small domestic biodiesel-production plant are life-changing in the sense that they demand regular input—you will have to put time aside for their operation and maintenance.

THINKING IT THROUGH

● **IS YOUR SETUP SUITABLE?**

Bioenergy systems all need a spacious and flexible setup, so you can only go ahead if your scenario perfectly fits the requirements of the system. For example, a composting toilet can only be installed if your home's layout is suitable, and the biodiesel plant can only be installed if your property fits the health and safety criteria.

● **DO COMPOSTING TOILETS SMELL?**

Yes and no. There is a chance that, in the early stages of the installation, your virgin composter might give off a whiff, but this will sort itself out once the system has settled down. Compost removal will also result in some earthy smells, but these are ones you can learn to love.

● **CAN I POUR KITCHEN SCRAPS INTO THE COMPOSTER?**

You can pour tea leaves or odds and ends of meals through the kitchen hatch in your composting toilet, but not great bucketfuls of liquid. The whole objective is to a achieve a nicely moldering heap of slightly moist, crumbly compost, so it does not make sense to pour lots of liquids down—it will hinder rather than help the overall composting process.

● **EVER-CHANGING SUPPLIES**

When it comes to woodchips and waste vegetable oil, remember that supplies can fluctuate. Don't proceed until you have thoroughly researched the situation in your area.

THINKING IT THROUGH (CONTINUED)

- **WOODCHIP STORAGE**

  Woodchips can be stored in hoppers, silos, cellars, and outbuildings—the choice will depend on your particular situation. For example, if you have a silo, are your local delivery people equipped to air-blow the chips from the truck up to the top of the silo, or will you have to install a pit with a screw-feed? You need to sort out all such issues before you settle on a storage option.

- **USE WASTE OIL ONLY**

  There are huge worldwide concerns about the implications of the growing number of oil-producing plants for biodiesel production. To us, it just does not make sense to take good farmland out of primary food production in a world where a good part of the population is going hungry. For this reason, we only advocate using waste oil for this purpose.

- **BIODIESEL GRANTS**

  Although from one country to another there are all sorts of grants and taxes that are designed to encourage and discourage various biodiesel options, remember that such "sticks and carrots" are liable to change.

- **WILL BIODIESEL DAMAGE MY CAR?**

  This is an ongoing issue. Make contact with biodiesel makers in your area and get advice. Most experts agree that the two main problem areas have to do with residues and the breakdown of natural rubber. Biodiesel will clean your engine, but the dissolved residues could possibly clog your filters; be prepared, in the first few days of biofuel usage, to change the various filters. As for the problems of natural rubber, the difficulty is that there is some evidence to suggest that biodiesel degrades some types of older natural rubber like washers, hoses, and gaskets. This is not a problem with newer cars, but if your car was made before 1994 then you need to change all natural rubber items to synthetic rubber.

Water energy

BASICS

Water basics

Water energy is the energy that we get from moving water and is not available to everyone in the same way that wind and solar energy are—you either have it or you don't. Flowing water sets a turbine or wheel in motion, and the turbine or wheel drives an electrical generator.

PICO-HYDRO TURBINES

Pico-hydro turbines are small-enough-to-carry versions of conventional micro-hydro–type water-driven turbines (see below). The smallest pico unit needs a constant water supply and a drop of about 5 ft. A unit of this size (depending upon the model) will only produce 200–500 watts—enough to provide basic power for a small cabin-type home.

MICRO-HYDRO TURBINES

These small water-driven electric turbines are a good off-grid option for remote farmers and homesteaders who have free access to plenty of water. Water is piped off from a point upriver, diverted through the body of the turbine where it generates electricity, and then piped out downriver. There are two options: a system that directly uses the generated power in the form of AC, and a system that produces DC electricity, stores it in a bank of batteries, and then puts it through an inverter.

WATER WHEELS

A small handmade "undershot" water wheel can be a good option if you have a shallow, fast-flowing stream. ("Undershot" means that the water hits the vertically placed wheel at the bottom, rather than flowing onto it from the top, which is known as "overshot.") Although this design is only about 25 percent efficient—much less efficient than an overshot wheel—it is relatively easy to build and install. The disadvantage is that such wheels are easily damaged by ice buildup.

FREQUENTLY ASKED QUESTIONS

- **Do I have the legal right to use the water that flows across my land?**
  Much depends upon where you live, but water rights have historically been the
  subject of many long-running disputes. You must establish your rights before you begin.

- **Can I connect my water turbine directly to my AC appliances?** If the
  water is running all the time, you can have an AC system (no batteries) that
  produces as much power as you want to use at any one time.

- **What is the best way of using DC power?** You can store it in a battery bank
  and use it, via an inverter, in the same way as on-grid household electricity; you can
  use it directly in the form of water heating; or you can use it to power DC
  appliances.

- **Do water turbines require a lot of heavy engineering and massive
  concrete groundwork?** It depends upon your setup and the size of the turbine.
  If you have the time and money, holding ponds, channels, control gates and so on
  are wonderful last-a-lifetime options, but in situations where it is not practical to
  construct channels, the water can be taken from upstream and pumped and jetted
  through a modern turbine. A turbine will not give you a gently turning picturesque
  water wheel, but it will give you power.

- **Are there any wildlife issues?** You have to make provision for fish to go up-
  and downstream, and you have to keep general noise and disturbance to a
  minimum. You cannot "muddy the waters."

- **Can I build a dam?** This will depend upon your situation. You can divert some
  insignificant part of the flow and build canals, holding basins, and so on, but you
  cannot delay, hold back, stop, or in any way disturb the main body of water.

What will it power?

Depending upon its size, a modern water-turbine system will produce as much electricity and heat as you want. A water wheel can also be rigged up to perform a mechanical task such as grinding flour. Your choice of system will be decided by the size of your stream or river and the way it relates to the shape and levels of your site. Water swiftly gushing from one level to another will call for one type of turbine, while a broad but fast-flowing river or stream will demand another.

ADVANTAGES AND DISADVANTAGES

● **ADVANTAGES**

- A water turbine in just the right location is more cost-effective than solar panels or a wind turbine.
- A small water turbine can be dedicated to giving you domestic hot water or to supplying heat through a resistive heater such as a storage heater.
- A small, low-power DC turbine in a fast/slow–flow scenario can be fitted with a controller and automatic switch and programmed so that it first charges your batteries and then heats the water.
- A water turbine can be used to power resistive loads, meaning items like immersion and convector heaters that contain a heating element.
- You can power all your "inductive" appliances such as a stereo, computer, television, DVD player, washing machine, and dryer.

● **DISADVANTAGES**

- You must have a good supply of water and you must have legal rights.
- The water source must be just right. The actual amount of energy will depend upon two factors—the rate of flow of the water and the height (the head) from which it is falling. In general, the larger the flow and the higher the head, the greater the power generated.
- In the context of power-resistive loads, you must make sure that the total power rating of the heating elements always exceeds the maximum supply potential of the turbine—otherwise the heating elements will burn out. While a good resistive option is to connect up to a water-cylinder immersion element that is always set in the "on" position, the downside is that this setup will use all available power from the turbine.

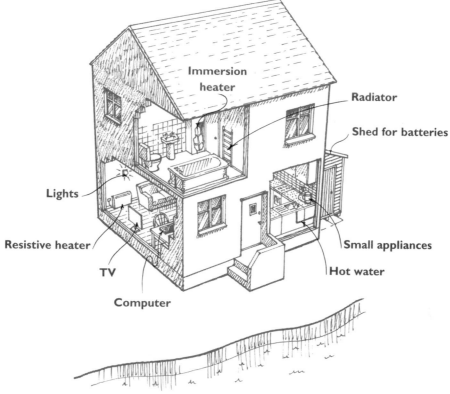

Immersion
heater

Radiator

Shed for batteries

Lights

Resistive heater

TV

Computer

Small appliances

Hot water

The range of uses for water energy

ADVANTAGES AND DISADVANTAGES (CONTINUED)

- While a large turbine is a good option for a remote location, you must be able to divert the water from a point slightly upstream, direct it through the turbine, and then return it to the river.

- While it is unlikely that all your appliances will be switched on at the same time, and in the knowledge that the turbine will only be charging the batteries when the water is flowing, you will need a battery bank large enough to take over when the turbine is "at rest" and a turbine large enough to supply all your requirements when the water is flowing. This can be very expensive. (To work out your power needs, see page 59.)

- Large turbines can be prohibitively expensive. If you cannot afford a turbine big enough to match your estimated needs, you will have to cut back. You could cut your power usage by economizing or supplement your system with another off-grid power source, such as another water turbine or photovoltaic cells.

PICO-HYDRO TURBINES

Pico-hydro turbine scenario

This is for you if you live in small cabin in a remote area, have a fast-running stream on the side of the property, have decided that you want to create power by means of a small pico turbine, and want to keep costs and groundwork to a minimum.

A remote cabin with its own turbine

QUESTIONS TO ASK YOURSELF

- Does the stream have, or can you create, a drop in level of about 5 ft.?

- Does the stream have a flow of 15 gallons per second?

- Can you manage on 500 watts of power—enough for lights and small appliances?

PROPERTY SIZE AND LOCATION

With this scenario, the size and location of your house—meaning the way it relates to the stream—and the shape, depth, width, and flow of the stream are all-important. As ever, with wind turbines and water turbines, the most efficient option is to set the system up so that there is the shortest possible distance between the turbine and the house.

HEAD OF WATER

A pico-hydro system requires a minimum head of water of 5 ft. There is no problem if your setup is like the illustration opposite—with lots of rocky outcroppings, waterfalls, and drops— but if you do not have or cannot create such a scenario, then this pico setup simply is not an option.

FLOW OF WATER

There are two miniature pico-hydro options: 200 W and 500 W. The 200 W model needs a flow of 9 gallons (35 liters) per second, and the 500 W model a flow of 18.5 gallons (70 liters).

POWER WITHOUT BATTERIES

If you are lucky enough to have a 24-hour, year-round flow of water running through the pico-hydro, then there is no need for expensive groundwork (channels and dams) nor for a bank of batteries.

CONCLUSION

If you are short on money, and/or if you want to use a stream without going to the trouble and expense of building groundworks, or if you feel in some way that your use of the water might be challenged, then a miniature pico-hydro pick-up-and-put-in-the-water system is a really good option. This system is so low in cost, however, that you could go for three units so that you have two in the water and one being maintained (giving you up to 1000 W).

Installing a portable pico-hydro turbine

Portable pico-hydro turbines are low in cost, relatively easy to set up and handle, and simple to maintain. If you have a fast-running stream and are looking for a quick and flexible system—one that you can have up and running in days and that can be removed just as speedily—then this is a good option. Another plus point is that it can be installed with minimum environmental impact. The disadvantage is that it will only produce a maximum of 500 W—enough to light five 100 W lightbulbs. If your water is running all year round, however, you can set up an AC supply without the need for batteries.

THE IDEAL SITE

Your stream must have a drop or fall of 3½ ft. A gushing stream that tumbles down in a series of drops is perfect, but at the other end of the spectrum you could have a setup where the water flows into a trough and then simply drops down through a tube or cowl that fits around the body of the propeller.

PLANNING THE LAYOUT

At the very least, you need a setup where the water is directed into a canal, so that it can drop through the body of the turbine. If you have a stream, there is plenty of water, and if you have a point where there is a dam and a drop of 3½ ft., you could mount the turbine by simply hanging it from chains or cables, and this will work in many situations. The best method, however, is to build a free-standing metal frame from angle iron and then set up the turbine and frame in much the same way as shown opposite. The idea of the frame is that the stability of the structure allows you to single-handedly make adjustments.

INSTALLATION TIPS

● If your stream does not give you a drop or fall of 3½ ft., then you can create one by building a weir or dam. If you are faced with such an option, you must make sure that the rise in water will not adversely affect your neighbors both up- and downstream.

● Although the alternator end of the turbine—the top end—is described as being waterproof, it is still not a good idea to have it under the water.

● After initially mounting the machine in a stable free-standing metal frame, you should, after spending time experimenting with various positions and setups, go on to build a more permanent masonry or concrete structure.

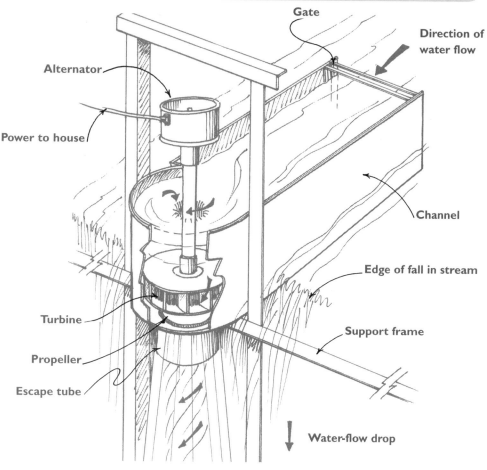

A portable pico-hydro turbine

MICRO-HYDRO TURBINES

Micro-hydro turbine scenario

Here, you live in the country in a hilly area, and you have a river or good-sized stream at a much higher level than your property. You have established your legal rights to use the water (with documented proof), and you have decided that you want to have an AC micro-hydro turbine (no batteries) to give you a total of 3 kW. Before you begin, talk to neighbors who have turbines, turbine manufacturers, and ground engineers. Learn as much as you can about what is available, and then make contact with a specialist and see what they advise. In some countries, there are grants and advisory schemes to help you.

QUESTIONS TO ASK YOURSELF

- The amount of energy available will directly depend upon water pressure and volume, so have you worked out pressure and volume figures?

- Your calculations should be based on 50 percent of the flow at the driest part of the year—say, half of the flow in the middle of a dry summer; so will your setup be suitable for your needs?

- You need to have a large fall of water over a short route—can this be managed on your site?

- Is the site you have in mind for the power house situated at a lower level than the water source, and yet is it in a position where the water can be run off back to the source?

PROPERTY SIZE AND LOCATION

Your property must be set close to a stream or river with the water at a much higher level than the turbine. If you can stand by the water and look downhill to what will be the site of the turbine, the chances are you have it about right.

THE HEADWORKS

Headworks are all the engineering works that go to make up the water intake, the point where the water is diverted from the source. Depending on your precise situation, you might need a holding pool with a weir or a holding pool with a gate or valve.

THE INTAKE

The intake is defined as the point at which the water is taken from the river or stream source. If there is a lot of debris in the source water, the intake and holding pool will need to include a settling pool, a place where the debris is separated out.

THE HEADRACE CANAL

The headrace canal, an open channel or pipe, directs the water from the intake pool, through another settling pool, and then into the penstock pipe.

THE PENSTOCK PIPE

The penstock pipe is a pipe that transports the water downhill and under pressure, from the tail end of the headrace canal and its associated settling tank to the turbine. This pipe must be as short as possible, be large in diameter, have few if any joints along its run, be smooth internally, and be made of a material that is mechanically fit for the purpose.

THE POWERHOUSE

The powerhouse houses the turbine, all its associated controls, and the upper end of the tailrace channel or pipe.

THE TURBINE

The turbine is the unit that converts the energy of the falling water into shaft motion that, in turn, by means of gears or belts, sets the electrical generator in motion. Your choice of turbine will depend upon your situation, your needs, the flow of water, and the pressure.

THE TAILRACE

The tailrace is the last channel or pipe in the system that sends the water on its way back to the river or stream.

CONCLUSION

The object of the groundwork is to take water from the river, remove debris, send it through the turbine and return it to the river, as swiftly and safely as possible, with minimum friction.

Installing a hillside micro-hydro turbine

A microturbine is very different from, say, a pico turbine or the traditional water wheel in that the source water is not only remote from turbine but at a much higher level. The water is diverted from the source, held in a tank or reservoir, passed downhill and under pressure through a pipe; it then passes on under pressure through a nozzle and through the turbine and then finally passed back into the source.

THE IDEAL HYDRO SITE

The ideal is to have the greatest fall of water over the shortest possible route. A fast-flowing stream running a short distance downhill and through the turbine, with power cables running to the house, is preferable to a setup with the water running a long distance downhill and across a valley to the turbine that is right next to the house, because the setup costs for the long pipe run will be huge. Remember that long runs of power cable are much cheaper than long runs of pipe.

SYSTEM OPTIONS

Whether you end up with a DC or AC system will depend upon your site and the flow of water. AC systems are good on many counts—they are able to supply much larger loads and you don't need inverters or banks of batteries—but they tend to be more costly in terms of the pipeline.

TURBINE SIZE

Turbine sizes range from small, low-cost, low-power DC models that are used to charge banks of batteries to large, expensive AC models at 20+ kW, and beyond.

HOW MUCH ENERGY?

Everything depends upon the volume and pressure of the water—meaning, in this context, the vertical distance between the water source and the turbine and the amount of water flowing between those two points.

IS A MICRO TURBINE A GOOD OPTION?

If you have a year-round supply of clean, fast-flowing water, then a microturbine system is perhaps the most efficient energy system there is. Certainly, the setup costs can be high, but microturbines are reliable, highly efficient, and generally low-maintenance.

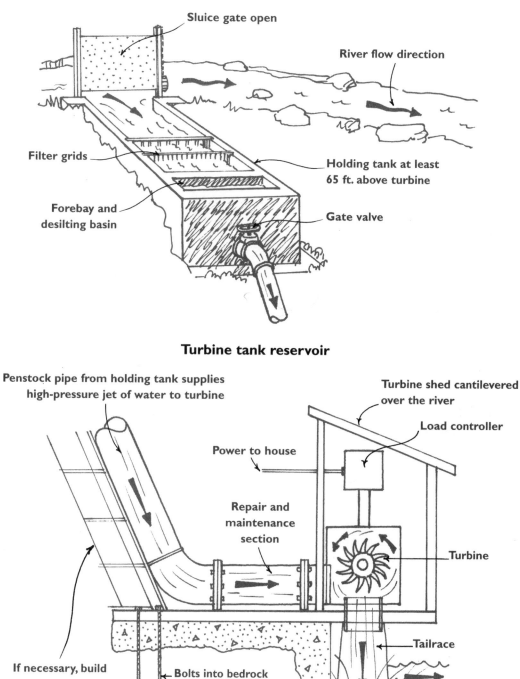

Turbine tank reservoir

Sluice gate open

River flow direction

Filter grids

Holding tank at least 65 ft. above turbine

Forebay and desilting basin

Gate valve

Penstock pipe from holding tank supplies high-pressure jet of water to turbine

Turbine shed cantilevered over the river

Load controller

Power to house

Repair and maintenance section

Turbine

If necessary, build a retaining wall to prevent soil slippage damaging machinery

Bolts into bedrock or foundation

High-grade reinforced concrete on bedrock

Tailrace

Cross-section of micro-hydro turbine

WATER WHEELS

Making a prototype water wheel

The idea with this project is that you are going to use your basic woodworking skills to build a prototype water wheel, so that once you have answered all relevant questions (such as best size of wheel, best number of vanes, rate of water flow, and best size and type of alternator), you will use this knowledge to create a more permanent setup. Further details are given on pages 132–33.

QUESTIONS TO ASK YOURSELF

- Your woodworking skills may be good, but what about the metalwork and the electrical? Will you need to bring in labor?

- Will you be able to run the water down a canal so that it catches the underside of the wheel?

- Is the location and character of your site such that you will eventually be able to build the canal from more permanent materials such as brick, stone, or concrete?

- Is the flow of the stream more or less constant, or is it intermittent to the extent that it dries up or floods?

- Do you have the legal right to cut into the water source?

PROPERTY SIZE AND LOCATION

As shown in the illustration on the opposite page, you will need a country property with a fast-running stream or spring on one side of your house.

THE WEIR AND CONTROL GATE

These function in much the same way as an on/off switch—you can shut the whole system down for repairs and maintenance. In the setup illustrated opposite, the weir and gate are built in an existing natural dip in the bank of the stream, but you might, if your bank is made from sand or mud, have to build a concrete structure.

Prototype water wheel (see page 133 for detailed drawing)

Waterside property with prototype water wheel

THE WHEEL

The wheel is 4 ft. in diameter, and made from a sheet of 1 in. thick, top-grade marine plywood. The two discs are set 2 ft. apart, with 12 vanes or buckets set at 30-degree intervals around the circumference of the disc, like the numbers on a clock face.

THE SHAFT, BEARINGS, AND PULLEY

The steel shaft needs to be about 1 in. in diameter, and long enough to pass through the wheel and extend well past each side of the bearings. A 2 ft. diameter pulley is used here.

THE CANAL

Since this is just a prototype, make the canal from whatever low-cost materials are on hand, such as plywood, steel, or glass fiber. It does not have to be completely watertight.

THE ALTERNATOR AND BATTERIES

The low-speed 15–30-amp alternator needs to be enclosed in some sort of shelter. If you have an existing brick or stone building in the right position, extend the shaft, run it into the building, and support the extended end with another pulley. Use deep-cycle batteries—say one 100 A.h. for the prototype and more for the permanent setup.

WARNING

A water wheel is a serious piece of equipment and is potentially dangerous. Under no circumstances allow children to use it as a playground.

Installing a DIY water wheel

Water wheels—the sort of things that we see in picture books and country guides—are the traditional means of converting moving water into mechanical power. The water turns the wheel, the wheel turns a shaft, and gears and cogs on the slow-turning shaft convert shaft motion into a mechanical up-and-down, in-and-out, or spinning motion. Here, the turning-shaft motion is used, through a pulley and belt, to drive the alternator.

THE IDEAL SITE

You will need a site where the water is running at a higher level than the surrounding ground, for example, a spring gushing from a cleft in a rock and then running down to a beach or a stream rushing down a broad valley. These are very different, but would both do the job. You must also have a situation where you can take the water off in such a way that it can be returned to the source. If your stream is just the right width and depth, with a rocky bed and banks, you might be able to set your system up so that it bridges the stream in such a way that the wheel can be lowered into the main flow. One ideal situation for a water wheel is shown on page 131.

MATERIALS

You can use wood, as we have done here, but there is no reason why you cannot work with metal, plastics, found machines, salvaged scaffolding, or whatever you like. Look at the overall design, and then modify and adapt the forms, structure, and materials to suit your likes, needs, and skills.

DESIGN

There are many variations on water-wheel design that work perfectly well. The design shown here is ideal for a location where the fast-flowing stream is running at a higher level than the house. Bearing in mind that every location is different, the big challenge for you will be how to reshape it to suit your needs.

SYSTEM FLEXIBILITY

One design problem is how to shape the structure so that you can experiment with a range of different channel angles. Lifting or lowering the channel is not so difficult, because you could have the bottom end supported on a beam or scaffold pipe that can itself be raised and lowered, but the difficulty is how to organize the alternator with its pulley wheel and belt so that it can be adjusted to follow the channel.

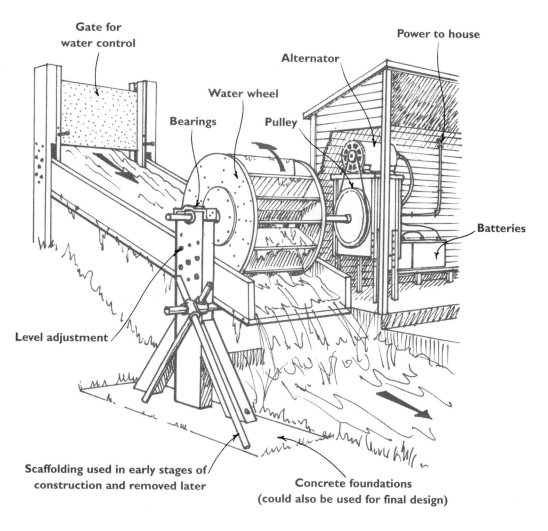

Gate for
water control

Power to house

Alternator

Water wheel

Bearings

Pulley

Batteries

Level adjustment

Scaffolding used in early stages of
construction and removed later

Concrete foundations
(could also be used for final design)

A simple water-wheel system

DEPTH OF WATER

The depth of water in the channel is critical in that, while its depth must be such that it pushes the wheel around, it must not be so deep that the surface of the water approaches the level of the center of spin of the wheel. Therefore, the height of the channel walls and/or the height of the wheel in the water might need to be adjustable.

CONCLUSION

If you have year-round water, and you get everything right, you will be able to produce year-round power. Once in place, a hydroelectric system is low-maintenance and long-lasting.

HINTS AND TIPS

Off-grid hydro-energy is something of a niche technology—after all, it can only be taken up by people who have running water—but one that is developing fast. Interestingly, the focus seems to be on creating systems that are ever smaller and greener.

THINKING IT THROUGH

- **THE WATER SOURCE**

  Research suggests that if you have a reliable year-round source of water, hydro energy is the best of all the off-grid, renewable-energy options. A hydro system— unlike wind or solar—is the only one that can provide round-the-clock power. The proviso is that the source must be reliable and constant.

- **YOUR LOCATION**

  Many properties have water, but of these only a few are practically suited to a hydro system. Swampy streams are no good, vast flowing rivers are just too much, canals are usually owned by remote companies, seawater lakes and estuaries usually have extreme tidal levels, and so on. It stands to reason that just about any location that comes near to being suitable for a traditional-type water wheel is one that has already been developed in the past. The good news for off-gridders is that hydro technology has shifted its focus from huge wheels to small turbines so that plenty of small streams are now possible contenders.

- **HOW MUCH WATER?**

  It is a fact that too little water is as big a problem as too much. The good news is that there is now a huge amount of interest in developing small, low-tech systems. Better yet, a mini–water wheel that can power a whole house from a water drop of about 8 in. has now been developed.

THINKING IT THROUGH (CONTINUED)

● **FIND THE RIGHT PROPERTY**

If you really want to power your life with a hydro system, you must seek out a property that has a stream, spring, or other similar water source. The disadvantage is that a year-round source of running water nearly always equates with a year-round rainy climate.

● **DO YOU NEED CITY/COUNTY PERMITS?**

Water is necessary for life, so it nearly always follows that, where there is water, there are long-standing laws and codes that govern its use. You can nearly always assume that you do need permits for anything other than a small standalone system. If, however, you are lucky enough to be living in, or moving to, a traditional house complete with its own water wheel, mill pond, and canals, it is pretty safe to assume that you can go ahead and use the equipment.

● **ADVANTAGES OF WATER POWER**

Hydro power is relatively clean and completely renewable; the electricity is cheap and pollution is minimal. If you get it right, the water coming out of your system will be as clean as the water going in.

● **DISADVANTAGES OF WATER POWER**

Large domestic systems where, say, a whole stream is being piped through a turbine can be disastrous to wildlife such as fish and amphibians.

Geothermal energy

BASICS

Geothermal basics

Geothermal energy is energy that we get from under the ground—from the Greek words "geo" meaning earth and "thermal" meaning heat. Geothermal systems take advantage of the fact that not far below the surface of the earth the temperature is a stable 45–58°F. A domestic geothermal system consists of a heat-transfer unit inside the building connected to a loop of pipes buried in the earth, dropped down a well, or layered and submersed in a deep pond or lake. A water/antifreeze mix is continuously circulated through the looped pipe and the heat-transfer unit. The liquid passing through the underground pipe gains heat, which is then extracted and directed to radiators, underfloor units, or forced-air units, or, if the system is reversed, heat is extracted from the house in order to cool it.

A HORIZONTAL CLOSED-LOOP SYSTEM

The continuous-looped pipe is buried in trenches or a pit at a depth of about 3–6 ft. This is a very good option if you have plenty of land and you don't mind having large areas of excavation.

A HORIZONTAL SLINKY CLOSED-LOOP SYSTEM

This option is much the same as described above. A greater length of pipe is needed, but the slinky, overlapping, loop-over-loop pattern of pipes requires a smaller excavation—you can fit more pipe in a smaller hole.

A VERTICAL CLOSED-LOOP SYSTEM

The continuous-looped pipe is dropped into a vertical borehole that is about 150–450 ft. deep so that the hole contains a single loop of pipe with a U-bend at the bottom. This is a more expensive option than a horizontal-trench system, but it does not need much space and uses less piping. If you are limited by space, or your land is rocky, swampy or in any way difficult to trench, then a borehole is a good option.

A POND CLOSED-LOOP SYSTEM

This is much the same as the horizontal closed-loop system, the only difference being that the looped pipe is 6–8 ft. underwater rather than underground.

A VERTICAL OPEN-LOOP SYSTEM

This system is like the closed-loop system in that the pipe is dropped down a deep borehole; however, here there are two boreholes. Ground water is piped up one pipe, pushed through the heat pump, and then sent back down the other borehole; or, if space and conditions allow, the exit water can be pumped over a field, or into a pond, stream, or river.

FREQUENTLY ASKED QUESTIONS

- **What is a geothermal heat pump?** A geothermal heat pump works like a refrigerator in that it extracts heat from one place—the refrigerator from food, a heat pump from the circulating fluid—and moves it to another. In winter, the geothermal pump extracts heat from the earth and sends it to your home; in summer, it extracts heat from your home and sends it to the earth.

- **Do geothermal systems need city/county permit?** Much depends upon the area and the planned system. A small, domestic, system completely on your property does not usually need permits, but always check with your local authorities first.

- **Can a geothermal system serve all my heating and cooling needs?** If you get it right, yes.

- **How efficient is the system?** Depending upon the system, you can plan on an efficiency of 300–500 percent, compared to, say, only 70–90 percent for coal.

- **What is the environmental impact?** Research suggests that, all things considered, geothermal systems are about as earth-friendly as you can get.

What will it power?

Geothermal energy is one of the cleanest, most efficient and most incontrovertible off-grid options. If you have the space and the money, and you are looking to heat/cool your home with a low-profile system, it is the perfect choice.

CONSIDERING THE OPTIONS

- If you have plenty of land, a horizontal closed-loop system (see page 138) is technically workable, but the required excavation is both messy and disruptive. If you have an open-field site, you can just bring in the tractors and bury the pipe, but it is not a good idea if your property has extensive established gardens or if the site is rocky, heavily wooded, wet, or sandy.

- A horizontal slinky closed-loop system (see page 138) is a good choice if your space is limited. This system concentrates the heat-transfer surface into a smaller volume, thus reducing the trenching by up to two-thirds.

- A vertical closed-loop system (see page 138) is more expensive than a horizontal-trench system, but if your property is not much wider than your house, or if your land is too difficult to trench, it might be the best or even the only way forward. Be warned that drilling costs can vary according to soil conditions, and you might be faced with an open-ended cost estimate.

- A pond or lake closed-loop system (see page 138) is only an option if you have a large body of water. A lake that covers ½ acre and is 7–8 ft. deep is about right for the average, well-insulated three- or four-bedroom home.

- While an open-loop system (see page 139) is a good choice in terms of cost-effectiveness and technical efficiency—especially if you want a single well to supply water for both the household and the heat pump—it might contaminate and/or deplete underlying aquifers.

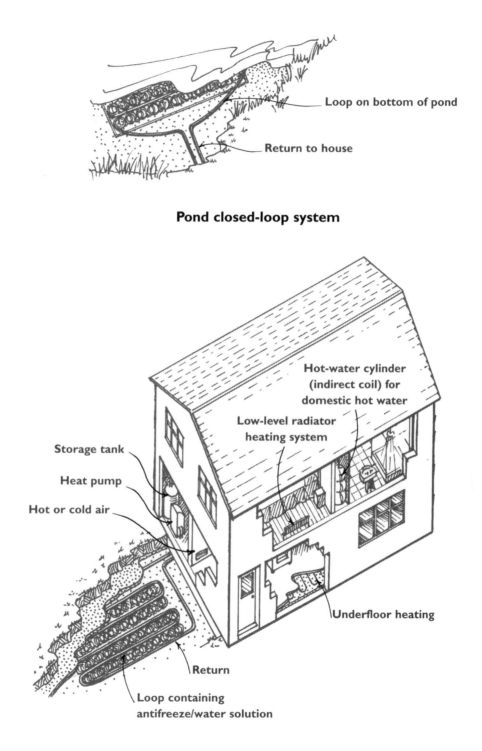

Pond closed-loop system

Horizontal closed-loop system

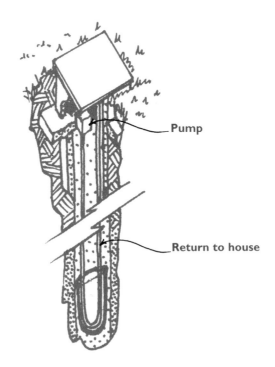

Vertical closed-loop system

ADVANTAGES

- A trench system is great if you have plenty of space, ideally if you have a large meadow on one side of the house.

- A closed-loop borehole system is a cost-effective choice if you want to kill two birds with one stone by sinking a single borehole to give you water for both the house and the heat pump.

- A closed loop in a pond is perfect if you have a good-sized lake on the property— probably the best of all eco-friendly, renewable, off-grid options.

- A heat exchanger rigged to a forced-air system is good in that you can have hot or cold air.

ADVANTAGES (CONTINUED)

- One of the best things about a geothermal system is that it is hidden away and quiet. This is very important if you live in an area where building codes are oppressive or neighbors are close by.

- A geothermal system can cut energy costs to the bone—you can save 70 percent on your household bills.

- A geothermal system has a long life expectancy of 30 years or more.

- You will be able to tailor a geothermal system to suit your particular site and your needs.

- The excavating techniques—digging holes and trenches—are relatively low-skill tasks that can be accomplished using local labor.

DISADVANTAGES

- A geothermal system can be expensive.

- Digging trenches and pits can be disruptive.

- Research suggests that heating efficiency relates directly to the insulation efficiency of your home.

- Digging a borehole is noisy, expensive, high-profile, and messy—perhaps not a good idea if you live in a highly populated area.

COUNTRY PROPERTY SCENARIO

In this scenario, you live in a smallish home on a large country property, you have plenty of land including a meadow, and you have decided that you want to install a geothermal system to give you heating. You have chosen a horizontal closed-loop system to keep costs down by using local labor to dig the trenches.

QUESTIONS TO ASK YOURSELF

- Do you own the land?

- Are you sure that the land is clear of all impediments such as underground water, gas or oil pipes, or power cables?

- Are you sure that the land is free from all restrictions—no rare species, endangered wildlife, archaeological ruins, special streams or ponds, rights of way, or protected soils?

- Do you need city/county permits?

- Will the gates, roads, and rights of way allow easy access for the workforce?

- Where in the house will you locate the heat exchanger?

PROPERTY SIZE AND LOCATION

The important thing here is the field. Once you have established ownership and rights, you can focus on all the details of the project.

THE TRENCHES

When using a large mechanical digger to dig trenches in wet, sandy, or rocky ground, there are only two choices: (1) you can dig part of the trench, put in part of the pipe work, and backfill as you go, in which case you will eventually be running the machine over trenches that have just been filled; (2) if space allows, you can dig one very long trench, put in the pipes, and finally fill. The second option is usually better because it allows you to see the various stages in detail.

An ideal property for a closed-loop system

A LARGE PIT

A large pit is a good option, but requires a great deal of space. For example, if you need to dig out a pit the size of a large swimming pool, you need space for the spoil—the murky slurry taken out of the hole—space for the machine to work, space for the pipes, and so on. If the ground is wet and swampy, or sticky clay, the site will be a quagmire. In this case, you would need to employ professionals.

STRAIGHT RUNS OF PIPE OR SLINKY?

For the plastic pipe work, there are three choices: straight runs, runs with occasional loops, or layered slinky loops. Slinky coils are expensive, but they require less land and shorter trenches, while long runs are expensive to dig and require more powerful pumps to push the liquid around.

HEATING AND COOLING

A geothermal system can provide both heating and cooling. In winter, the fluid in the pipes absorbs the heat from the earth and gives heat to the house; in summer, the pipes take heat from the house and give it to the earth. In both setups, the system provides hot water.

CONCLUSION

A geothermal system saves on energy, is simple to operate (no smells, noise, fumes, or flames), is safe, and the heating/cooling is evenly distributed. Geothermal fuel costs about one-third that of natural gas and about a quarter that of liquefied petroleum (LP) gas. The disadvantages are the need for ground space and the high setup costs.

Installing a horizontal closed-loop system

If you have the land, and if access is easy, a horizontal closed-loop system is one of the most cost-effective of all the geothermal options. At first glance, the notion of a "horizontal closed loop" might sound a little daunting and complicated, but it is just a plastic pipe that goes from the heat exchanger around a pattern of trenches and back to the heat exchanger. The fluid in the pipe is pumped from the heat exchanger, around the buried loop, back to the exchanger, and so on—around and around.

COST AND EFFICIENCY

With geothermal earth-loop technology now being completely understood and relatively commonplace (component parts can now be purchased in stores), the setup costs and subsequent efficiency of the system will depend upon your choice of earth-loop configuration and the cost of the trench work. For example, if your land is restricted in some way, you could have short trenches with slinky runs of pipes (as shown here), or, if you own a large area of land, you could have long trench runs with straight pipes.

THE TRENCHES

Your pattern of trenches—the length, layout, depth, and configuration—will depend upon your site. You may have a site where you can easily clear ½ acre to a depth of 3–6 ft. and still have space all around; you may have a site where the trench needs to snake around existing features such as trees, buildings, or underground sewage systems; or you may have a long, narrow site where the only option is to have a single long trench traveling down the yard. Therefore, it is almost impossible to say precisely how long your trenches and pipes need to be, except that a typical horizontal loop will have 400–600 ft. of pipe.

USING A BACKHOE

When using a backhoe to excavate a trench, the procedure will depend upon space. In some cases, the backhoe would have to come in the front entrance and travel down the yard with the machine bridging the trench and the spoil being put to the sides; then it would either come back up the same route filling in the hole, or it would go out of a bottom gate, travel around on a public road, come in the entrance and start again. The procedures and costs will always depend upon the complexities of the site.

THE SLINKY LAYOUT

Pipes are now usually relatively cheaper than digging trenches and are consequently becoming more popular. The introduction of the slinky pattern means that you can have a much greater length of pipe work in the same area of ground. The trench must be wide enough to allow the slinky coils to settle down, and when the coils are in place and the system has been pressure-tested, the trench must be back-filled in such a way that the pipe work is not damaged by sharp rocks, debris, or impact.

CONCLUSION

If you are hiring a backhoe, the simplest digging option is a single, long trench, but if you are bringing in labor you should always take the professionals' advice.

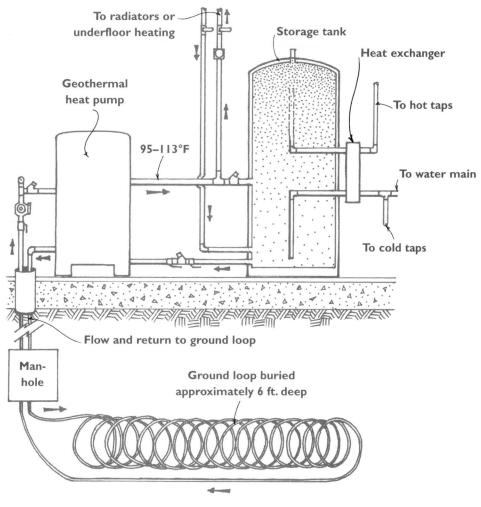

Geothermal ground loop installation

NARROW COUNTRY SITE SCENARIO

In this situation, you live in the country but your usable ground space is restricted by structures, trees, and rocky, rising ground all around. You have decided that the best solution is to install a geothermal system with closed loop set in a borehole.

QUESTIONS TO ASK YOURSELF

- Do you own the land, and are you sure that the land is clear of all impediments such as underground water, gas or oil pipes, or power cables?

- Do you need city/county permits?

- Will the gates, roads, and rights of way allow easy access for the workforce?

- Where will you site the borehole?

PROPERTY SIZE AND LOCATION

If you are short of ground space around your house, your only real geothermal choice is a vertical closed-loop system in a borehole.

BOREHOLES AND DRILLING

The holes are not actually drilled, but are "driven." Cable-percussion "drilling" is just about the oldest, simplest, and most low-tech of all the drilling techniques. A rig arrives on site on a trailer or truck, a tripod is erected complete with an engine, metal cable, and various tools, the machine is set up, and then the engine is set in motion. A large, bullet-shaped tool is pulled up and dropped hundreds of times. As the tool drops, it mashes the ground at the point of impact, creating an ever deeper hole. Every now and then, a tube-shaped bucket is lowered into the hole to bail out a mix of water and crushed material. The diameter of the hole is governed by the width of the bullet-shaped tool, and the depth is determined by the length of the cable and the number of times the tool is dropped. Some of these machines are so small, and/or break down into such small component parts, that they can be manhandled

Property with restricted space

across difficult terrain, through forests, and even through doorways. The spoil can either be taken away in a tanker, or, if the space and situation allow, emptied into a dip in the land.

WATER IN THE BOREHOLE

A water "strike"—meaning finding good quantities of quality water—will be indicated by the color and consistency of the bailed slurry. If you are lucky, and the drilling is easy and the water plentiful, it might be possible to drop a dual system down the well—a closed-loop pipe for the geothermal heating and an open pipe for domestic water for washing.

BRINE OR ANTIFREEZE IN THE LOOP

The vertical loop is inserted into the borehole so that the chosen fluid medium runs deep into the ground and back again, bringing up heat from the surrounding earth. Some systems use brine, and others a water/antifreeze mix. If you are going to use a single borehole for a closed-loop system and a household water supply—and this is only possible in the context of a borehole with a high water yield—then use a brine solution in the closed loop rather than potentially dangerous antifreeze.

THE CLOSED-LOOP PUMP

The fluid in the closed loop has to be pushed around the system—around a great length of pipe—by a pump fitted on the outside of the heat exchanger. When conditions are such that you have to increase the length of pipe, use a larger diameter of pipe, or have a convoluted pattern of pipe, it may be necessary to install an additional pump.

Installing a vertical-loop system

If your land is difficult—perhaps rocky, covered in trees, studded with structures, or undulating—the best solution is a vertical-loop system in a borehole. The system can be either open or closed.

OPEN- AND CLOSED-LOOP SYSTEMS

The open-loop system (where the pipe is actually open-ended rather than looped) is a good choice if you have a borehole that gives you unending supplies of water. The water is pumped from the borehole to your home, heat from the water is transferred to the heat pump, and finally the water is pumped out of the building and dumped into a suitable drainage area such as a field, pond, lake, river, stream, or another borehole. The vertical closed-loop system works in the same way as a horizontal closed-loop system, except the loop is dropped down a borehole rather than spread out in trenches.

WATER VOLUME AND QUALITY FOR AN OPEN-LOOP SYSTEM

For an open-loop system, while you don't need a great deal of water, the supply needs to be constant and smooth. Pump the water out over a given period and measure the throughput to give you a figure in gallons per minute. It is no good if your pump is sucking at an unsteady supply. You also need to test the water for iron, acidity, hardness, and organic matter. Once armed with this data, make contact with several heat-pump suppliers to see if they have an exchanger to match up to your figures—the right size for the flow, and the right design and material make-up for the water.

IS AN OPEN-LOOP SYSTEM ENVIRONMENTALLY FRIENDLY?

There are three potential problems with open-loop systems: how much water you should take, where should you put the water once you have finished with it, and whether there is a chance that somehow you will introduce toxic or otherwise damaging materials into the system. Research suggests that there will be little or no change to the quality of the dumped water other than a slight temperature decrease or increase. There is always the danger that your system will damage river banks or stir up sediments. Contact your local authority before you start to check for any codes and restrictions.

USING A CLOSED-LOOP SYSTEM

A closed loop in a borehole is a good option where you cannot dig trenches for any reason or where you can have a borehole but are not allowed to dump water. The depth of the borehole will depend on the location of your site, but most holes are 200–600 ft. deep. A borehole system is expensive, but it will require fewer work hours than a horizontal loop to put in place and considerably less pipe work to get it up and running. The closed loop will need to travel from the heat exchanger, through a pump, under the ground to the well head, down the borehole, into one side of a preformed U-bend unit, out the other side, up the borehole to the well head, and back underground to the heat exchanger. This should be straightforward if you spend time planning out the route of the pipe.

CONCLUSION

The big concern with a borehole system is, of course, the hole—the expense, its precise location, and all the issues relevant to bringing in machinery. Spend a lot of time at an early stage visualizing the whole project, and try to iron out all the wrinkles at the design stage.

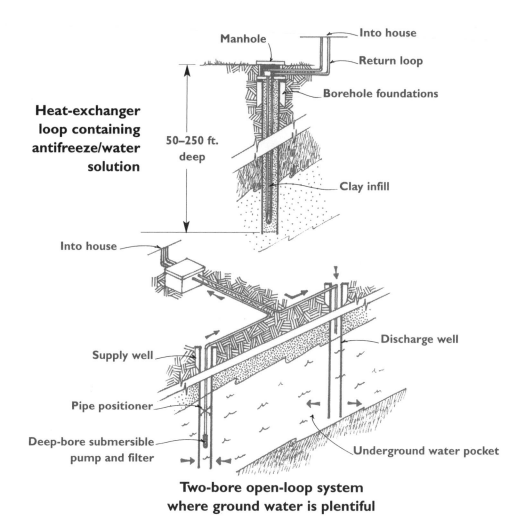

**Two-bore open-loop system
where ground water is plentiful**

LARGE PROPERTY WITH A LAKE SCENARIO

For this scenario, you live in the country on a largish plot where most of the ground space is taken up by a large natural lake. There are structures, trees, and rising ground all around. You have decided that you want a geothermal system with a closed loop set in the lake.

QUESTIONS TO ASK YOURSELF

- Do you own the land and the lake, and are you sure that you have legal rights to use the lake?

- Do you need city/county permits?

- Will the gates, surfaces, roads, and rights of way allow easy access for the workforce so that they can get right up to the water?

- Precisely where on the lake's edge will you run the pipes into the water?

PROPERTY SIZE AND LOCATION

Because just about everything about this project is focused on the lake—the depth of the water, the quality of the water, the geology of the bed of the lake and its surroundings, the way that the land lies between the lake and the house, and so on—you must make sure that you have clear rights regarding the lake, the water in the lake, and all the land between the lake and the house. For example, there could be a restrictive covenant in the property deeds that restricts the use of the water in the lake—meaning that building permanent structures such as a geothermal closed loop would be prohibited. Always double check your rights before you start.

IDEAL LAKE DIMENSIONS

The lake must be at least 8 ft. deep for a good part of its area. Research suggests that for a typical three- or four-bedroom house you will need a body of water with a surface area of about ½ acre.

Property beside a large lake

PUTTING THE LOOPS IN PLACE

Once you have established the best size of pipe and the best slinky-coil layout for your area of water, then comes the tricky business of actually getting the pipe into the water. The best method is carefully to float and layer the coils into place, tie them together to make a raft, and then tie or strap weights at various points both on and around the raft so that it is fixed in place and slightly weighted. Once the raft of pipes is in place and all joints and manifold connections have been pressure-tested, the pipes are slowly filled with fluid so that they sink slowly to the bottom of the lake.

HOW LONG WILL THE LOOP LAST?

Research suggests that, if the pipes are sound and if they are left alone (no boat hooks, sharp anchors, heavy weights, or slashing boat motors), they will last indefinitely.

ENVIRONMENTAL CONSIDERATIONS

The high-density polyethylene pipes must be thermally butt- or socket-jointed according to various codes and standards—there should be no glues or mechanical joints.

Installing a closed-loop system in a lake

If you have a good-sized lake (see page 152), you will be able to install the best of all geothermal options. The ideal is a lake about 10 ft. deep, where the banks run gently into the water and the bottom is level and sandy.

GETTING THE PIPE INTO THE WATER

Installing a slinky layout for a closed loop in a lake is similar to installing a horizontal closed loop in trenches, but the difference is that you will be working in the water with rubber boots, small boats, and snaking ropes rather than moving earth with diggers. The first thing to understand is that, although the pipe is relatively heavy and unmanageable on land, it floats when empty and is more or less suspended in the water when it is full of fluid. The best method is to build a metal framework or containment on the bank, as close as possible to the water, and float the pipe into place.

To do this, position concrete blocks on wooden skids at, say, 3½ ft. intervals, build a metal framework on the blocks, arrange the layered slinky loops of pipe on the frame, and strap the frame to the blocks, and the pipe to the frame, with plastic tags. The frame could be made of scaffolding pipe held together with clamps, or it could be bolted and/or welded, as long as it is strong, resistant to corrosion, and free from sharp corners and edges. Once the raft of pipe work is ready, and you have spent time painstakingly lashing and strapping the slinky loops to each other and to the framework, drag it into the water and very gently float it into position.

Finally, when you are happy with the raft's position, you have connected it up to the pipes that run underground from the house to the lake, you have pressure-tested for leaks, and you are sure that there is enough slack in the linking umbilical pipes to allow for the depth of the water, fill the whole loop with the antifreeze liquid. At this point, the whole thing will gently sink into the water.

CONCLUSION

The big issue with this system is how to construct the raft in such a way that it will bear the out-of-water weight of the pipe, stand up to the strain of being dragged into the water, and then be able to just sit there in the water without rusting or at least without damaging the pipe work. There are just about as many raft-frame options as there are scenarios—some installers build structures so that the looped pipe is contained like plates in a rack, others build lobster pot–like containments from galvanized tube and mesh fencing, and yet others simply lash the coils of pipe together and drop them in the lake. The one shown here is, in our opinion, the most efficient all-around system.

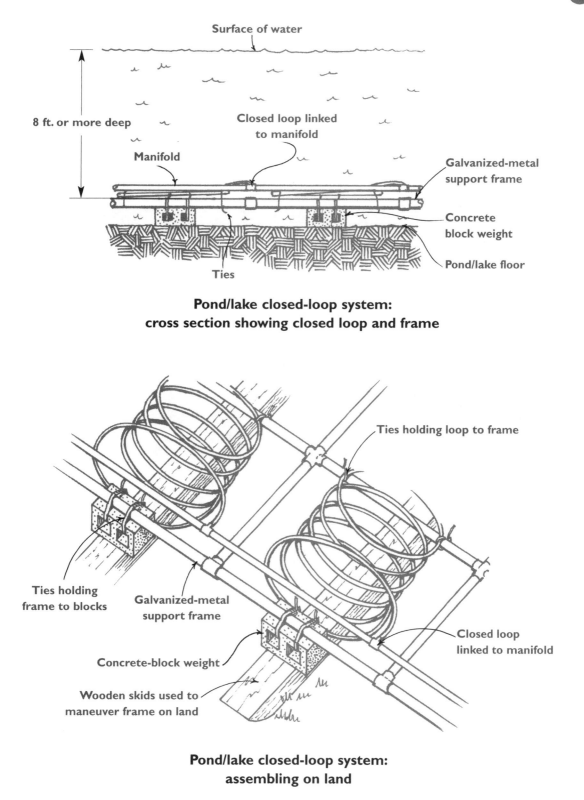

Pond/lake closed-loop system:
cross section showing closed loop and frame

Pond/lake closed-loop system:
assembling on land

HINTS AND TIPS

It is slightly puzzling that the potential of small geothermal heating systems has been known about since the 1940s, but it is only now that they are becoming popular. It is predicted that domestic geothermal systems will be one of the foremost innovations of the twenty-first century.

THINKING IT THROUGH

- **ARE HEAT PUMPS GOOD OPTIONS ON AN OLD PROPERTY?**
  Heat pumps, sometimes called heat exchangers or geothermal heat pumps (GHPs), work like refrigerators in that they extract heat from one place and move it to another. If you live on an old property—perhaps in a damp and badly insulated house, where it gets very hot in the summer and icy in winter—then a heat pump would be working desperately against the odds, to the extent that it would be a bad option. The best thing to do would be to insulate the building to the maximum and then go back to the geothermal idea.

- **CHANGING AN EXISTING HEATING SYSTEM**
  Most people are living in homes with traditional heating systems, so it does not make much sense to take out the system and install geothermal. However, if the prices of fossil fuels go any higher and if supplies become even more tricky, it will make good sense to cut your losses and start fresh.

- **WHICH LOOP SYSTEM IS BEST?**
  There are five main loop options: horizontal closed loops in trenches; horizontal closed loops in large, field-size excavations; vertical closed loops in boreholes; open loops in boreholes and lakes; and closed loops in lakes. There are also mixed systems. Most people have only one option that suits their site.

- **WILL A BOREHOLE BE EASY TO CREATE?**
  Creating a borehole used to be costly and difficult, but machines are now smaller and more efficient. When you come to agreeing on a price with your contractor, make sure you agree on a fixed price for a fixed depth or hole.

THINKING IT THROUGH (CONTINUED)

- ### HOW MUCH WATER DOES AN OPEN-LOOP SYSTEM NEED?

  Your system will need about 5–10 gallons per minute. Once you have used it, the water has to be discharged into a stream, river, or other place. You need to know a few hard facts. Do you need permission from local/national authorities? Will you be able to extract a constant year-round supply from your borehole? Can you discharge the water into your chosen place? Will the force of the discharging water in any way adversely impact upon the environment?

- ### OPEN LOOPS AND WATER QUALITY

  Does your water contain anything that will damage the pumps in the loop and the heat exchanger, such as iron, sand, plant life, calcium, and will such contents damage wildlife at the discharge point? Have the water tested, and make sure your pumps and heat exchanger are up to the task and that you can safely discharge the water.

- ### LAWS, LOCAL CODES, AND RIVER AUTHORITIES

  The whole subject of water is fast becoming a highly contentious issue. If you are thinking of extracting and discharging relatively large quantities, always check with all local laws, codes, river authorities, nature-protection societies, and so on, and get them to give you signed and sealed documentation.

- ### WHAT IS AN AQUIFER?

  An aquifer is an underground body of water where water is stored in porous rock—a little like water in a bath sponge. Aquifers can be as small as a village or as big as a country. When you have a deep borehole that gives you a lot of water, you will be taking it from an aquifer. The problem with using such water—pumping it out and maybe even pumping it back in—is that you might in some way be introducing contaminating materials such as chemicals, oil, animal and vegetable matter, or sewage. You must ensure that your system is 100 percent clean.

- ### ARE HORIZONTAL GROUND LOOPS TRICKY TO INSTALL?

  No, but the whole business of bringing all the ingredients together requires a lot of time and energy. It is difficult because there will need to be sizable excavations that must travel under, over, or around existing features. The real challenge with a project of this character is how to do it all with the minimum of disruption.

THINKING IT THROUGH (CONTINUED)

● **BACKHOES**

A backhoe is a wonderful machine, but there is no getting away from the fact that it is large and heavy and will make a mess. You will need to do as much preparation for the work ahead of time as possible in order to minimize the disruption—for example, cut back trees and relocate plants.

● **SPACE REQUIRED FOR DIGGING TRENCHES**

If you are digging your own trench with a spade, simply start at one end, take out a "spit" (one spade's depth) of earth and put it down on one side (depending upon whether you are left- or right-handed), and gradually work backward. At the end, you will have a trench with the spoil on one side, and a route for walking on the other. In this way, a simple trench about 2 ft. wide and 6 ft. deep will take up a strip of land about 8–15 ft. wide. So it is with a machine—if you are digging a trench 6 ft. wide and 6 ft. deep, you might take up a strip of land about 15–20 ft. wide. Of course, you can dig a section of trench, set the pipes in place, and fill as you go, but it will not be easy.

● **SOIL CONDITIONS**

Depending on where you live, your land might be anything from solid rock to loose gravel, sand, clay, deep loam, and so on. Your land might be wet, dry, level, sloping, and full of cavities. If we assume that at least three factors make-up the character of your land—the dips and slopes, the structure of the soil, and the amount of water in the soil—you can see that digging trenches can be anything from easy to tricky. You may need to shore up the sides of the trench if the soil is sandy or use a pump to clear the water if you encounter clay. Clay is particularly difficult in that it takes time to settle back in place; there will be a raised lump running along the line of the trench that might take a year or so to level out.

● **WHAT SORT OF PIPES TO USE**

Currently, most experts agree that the best material is "high-density polyethylene." Research suggests that pipe of this character buried in the ground will last up to 200 years. Most companies will give you a 50-year guarantee. As ever, be warned that there are companies who will try to cheat you by selling inferior-quality pipe.

THINKING IT THROUGH (CONTINUED)

- **PIPE JOINTS**

  The joints are all the U-bends, T-junctions, and straight sockets that join lengths of straight pipe to make up the total layout. To date, the only proven method of jointing pipes is a technique known as "thermal fusion" where the pipes are melted together. There are two ways to make a thermal joint: (1) the ends of the pipe are trimmed square, pushed together, and heat-fused; (2) the ends of the pipe are trimmed square, pushed into a socket piece, and heat-fused. Both techniques use special tools and systems. In case you are tempted to cut costs, on no account should the joints be glued and/or held together with mechanical clamps.

- **HOW CAN I TELL IF THE JOINTS ARE SOUND?**

  The piping is going to be more or less inaccessible, so you must test the system. Fill up the system with water under pressure from the water main and leave for a day or two for the pipes to stretch. Once they have stretched fully, fill up again and test the water pressure. The gauge will give you a positive reading if there are problems.

- **WILL I NEED A SPECIAL ELECTRICITY SUPPLY?**

  The system needs electrical power to set it in motion, but this is usually taken care of by a normal domestic system. If you are completely off-grid, you will need a generator, a good-sized bank of PV cells, or a wind turbine.

- **HEATING THE HOME**

  The heat from a geothermal system can be used variously as radiant heat that is delivered through underfloor pipes or low-level radiators, as hot air that is delivered through ducts, and as hot water for domestic use. A heat pump can provide all of your domestic heating and cooling needs for little or no cost.

- **SITING THE GEOTHERMAL HEAT PUMP**

  Most systems are located indoors rather like traditional water heaters. The good news is that most modern heat pumps take up less space than conventional boilers.

- **CLEAN AIR**

  An electrically operated geothermal system is totally clean—no dust, gases, noxious smells, carbon monoxide, soot, or residues.

Gas energy

BASICS

Gas basics

In the 1960s Harold Bates, a chicken farmer in rural England, produced a methane-powered system that he used to fuel his car. A mix of manure, chopped straw, and water went in at one end of a digester, the mix was heated and allowed to decompose under pressure, and usable methane gas came out at the other end. If you have large quantities of organic waste (pig manure, cow manure, chicken manure, human manure, chopped-up waste food, for example) and you are able to control the throughput somehow, then a methane-making digester is a real possibility.

FREQUENTLY ASKED QUESTIONS

- **What kind of manure do I need?** Bates worked out that a mix of 40 percent pig manure, 35 percent chicken manure, and 25 percent chopped straw, all nicely doused with water, produced usable quantities of methane. Some mixtures work better than others, but just about any organic materials that can be anaerobically decomposed can be used to produce biogas.

- **Can I turn my septic tank into a digester?** You can, but the process will not be thorough enough to produce usable amounts of gas. If you can stop putting chemicals such as bleach down the drain, however, and maybe add additional animal manures to the human waste, then it should work.

- **Does a cold climate affect production?** Bates believed that the secret of success was to heat the mix to a temperature of about 85°F. If the temperature of the sludge is much higher or much lower than this, the process does not work. This being so, it follows that efficiency is likely to be higher in a warm climate—everything will be faster.

- **What can I do with the exhausted sludge?** Once the organic mix has produced gas, the sludge can be removed and used on the vegetable plot, just like compost. It is a wonderful growing medium.

FREQUENTLY ASKED QUESTIONS (CONTINUED)

- **Is methane-gas production a workable option for a family home?** The simple answer is no. Research suggests that small digesters are most efficient when they are built to serve a community of several families, a small town, a group of farms, or a large pig or dairy farm—a situation where the manure-to-methane ratio is high. One such community, a group in Santa Fe, New Mexico, has built a digester that they keep filled up with cow manure. This community has the ideal setup: lots of manure, warm temperatures, very few cleaning agents such as soap and bleach, and relatively low fuel needs. They use the gas mainly for cooking.

- **How much of the gas produced is methane?** In general terms, a mix of human and animal waste will produce about 60 percent methane and 40 percent carbon dioxide.

- **Is methane environmentally friendly?** Because a digester produces 60 percent methane and 40 percent carbon dioxide, with no sulphur emissions, and the waste can be used on the vegetable garden, research suggests that biogas production is all good. Certainly the process produces carbon dioxide, a greenhouse gas, but this is no more than would be produced in the normal course of events. Methane is also a greenhouse gas, but in this system it is put to good use rather than released into the atmosphere.

- **What can methane be used for?** Methane can be used for heating water, gas lighting, running a generator, fueling a car, and cooking.

- **WARNING:** Biogas is both explosive and toxic, so it must be treated with extreme caution at all times. In some countries, you are required by law to be a qualified gas engineer before you can work on installations of this character.

What will it power?

Biogas is a mixture of methane and carbon dioxide produced by the anaerobic digestion of organic wastes. The active part—methane—can be used for cooking, heating, running engines, lighting, and so on. In a domestic system, gas production is managed in a "digester." These come in many shapes and sizes, but all have similar design features and functions.

ADVANTAGES

- It is possible to build a DIY system using easy-to-find, basic materials.

- The system renders potentially dangerous materials such as pig manure harmless—the digesting process kills pathogenic materials.

- High-quality fertilizer is a byproduct of a biogas plant.

- If you have an off-grid setup that gives you lots of animal manure, it is possible to build a neat circular system: Animals produce manure, manure is turned into gas and fertilizer, fertilizer is used on the garden, gas is used for lighting, cooking, and heating, heat and light are used for the animals' welfare, and so on.

DISADVANTAGES

- The chemical reactions are such that metals (except iron and nickel) will kill off the anaerobic process. Therefore you must use polypipe-type fittings—no brass or copper. The gases will also corrode any metal parts.

- You need unlimited supplies of raw material such as animal manure or some kind of food production waste, which can be smelly and unpleasant to handle.

- Components in the biogas, such as ammonia and hydrogen sulphide, need to be removed by being "scrubbed" (see page 167).

- Biogas is explosive when mixed with air, so you cannot use electric power tools or anything that might ignite the gas. Methane can also render you unconscious.

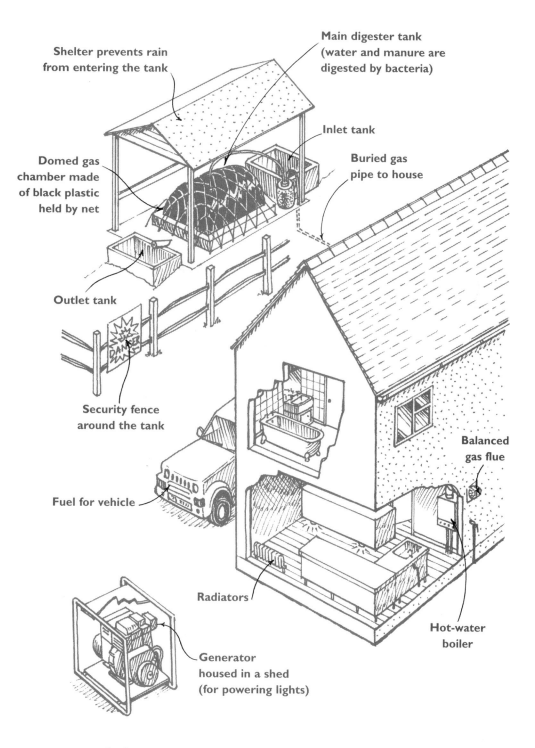

Shelter prevents rain
from entering the tank

Main digester tank
(water and manure are
digested by bacteria)

Inlet tank

Domed gas
chamber made
of black plastic
held by net

Buried gas
pipe to house

Outlet tank

Security fence
around the tank

Balanced
gas flue

Fuel for vehicle

Radiators

Hot-water
boiler

Generator
housed in a shed
(for powering lights)

A domestic biogas system and its range of uses

SMALL-COUNTRY-FARM SCENARIO

If you have a completely off-grid small farm, with geese, chickens, and pigs in an isolated spot on the edge of a remote, small-town community, you may want to install a digester to provide you with power. This project requires gas-installing skills. It is best to start with a prototype system and later build something more permanent and sophisticated.

QUESTIONS TO ASK YOURSELF

- Gas is potentially a killer, so do you have the required skills and knowledge, or should you employ a professional?

- Do you have suitable tools, such as bronze grips?

- Do your animals and/or birds produce enough manure—about 5 gallons a day?

- Are you up to the task of digging a hole, laying concrete blocks, and mixing concrete and mortar?

PROPERTY SIZE AND LOCATION

This system will involve you physically moving manure from the source to the digester, piping gas from the digester to the house, and moving spent slurry from the digester to the garden, so you must find a location that will keep all this heavy bucket-moving to a minimum.

THE MAIN DIGESTER TANK

Inside this warm, underground tank—covered with a domed gas-tight lid or chamber—the bacteria in the manure thrive and produce biogas. For this tank, you will need to create a hole 6 ft. deep, 6 ft. wide, and 9 ft. long. These dimensions are based on an initial fill-up, then a daily input of about 5 gallons of water and 5 gallons of manure. For further details, see pages 168–69.

THE INLET TANK

A tank, pit, or container, usually fitted with a mixing device, is filled with the manure mix. When the mix has been well stirred, a plug, sluice, valve, or gate is opened and the charge of slurry slides down into the main digester tank (see illustration at the top of page 165). The inlet tank needs to be big enough to take 10 gallons of liquid. It should be placed well above the top of the digester with the feed pipe running right down to the bottom.

THE OUTLET TANK

With a continuous-feed digester, the action of an amount of fresh slurry being fed into the inlet tank pushes the same amount of exhausted slurry into the outlet tank. This must be at least as big as the inlet tank, but at a slightly lower level, with the outlet pipe more or less coming up from the top of the main chamber. The daily input of 10 gallons will displace the same amount—in at one end and out at the other.

THE GAS CHAMBER AND OUTLET

The gas bubbles up through the slurry and collects in the domed chamber above the main digester tank. Depending on the design, the chamber will be either fixed or inflatable. The dome is like an upsidedown mixing bowl floating on a water-jacket gasket. A line of pegs sticks out from the inside walls of the pit to prevent it from sinking below a certain level, and a rope mesh stops it from floating away. There will probably be some gas leakage, but this can be minimized by making sure that the dome is airtight, and the water-seal space between the dome and the sides of the pit is kept filled up. When the gas reaches a predetermined pressure, it is passed through a scrubber (see below) and either piped directly to the house or stored in containers for later use.

LIQUID LEVELS

The liquid level within the system, while well below the level of the inlet pipe, is only slightly below the top of the outlet pipe. The levels are important because they allow the daily input down the input pipe to push the same amount out from the exit pipe.

THE "SCRUBBER"

The function of the scrubber (see illustration on page 169) is to filter and clean out the impurities from the gas, mainly hydrogen sulphide, which if unchecked will combine with the moisture in the gas to create sulphurous acid that will in turn corrode just about everything. The scrubber is a large glass container stuffed with wire wool. Gas goes into the scrubber, the sulphurous acids attack the wire wool, and clean gas comes out the other end.

Making a biogas digester prototype

If you have limited funds, enjoy DIY, and are not afraid of a little hard work, this project is for you, but it does require a qualified gas engineer. Remember that biogas is both explosive and toxic.

MATERIALS

You will need: enough bricks, concrete blocks, or concrete to line the pit (see below); sand, cement, and aggregate; a sheet of best-quality plastic sheet or butyl rubber about 20 ft. long and 9 ft. wide; about 15 ft. of 3-in. diameter PVC tubing; about 30 ft. of ½ in. diameter PVC pipe for the dome, and enough to run as far as the house; nonmetal fittings for the pipe; and all the other bits and pieces that most people either have or can get. These quantities are just a guide—if you are clever and qualified enough to go ahead, you will probably use your own plans, designs, and modifications. Start by building the basic pit, and then buy the other materials to suit the finished size of the pit and your specific house-to-digester needs.

CREATING THE DIGESTER TANK

Dig out a pit about 7 ft. deep, 8 ft. wide, and 12 ft. long with trenches running in at each end for the inlet and outlet pipes. Pour a 6 in. thick layer of concrete into the bottom of the hole and tamp it level. Wait for the concrete to cure, and then build the walls of the digester up from the base. Aim for finished inside dimensions of about 6 ft. deep, 6 ft. wide, and 9 ft. long. Don't worry if the sizes of your blocks or bricks guide you toward a hole of slightly different dimensions, and don't forget to leave holes in the end walls for the inlet and outlet pipes.

Once the walls are built to three-quarters of their final height and the mortar has cured, mix concrete and pour it in between the walls and the soil. If your soil is sandy, muddy, and/or in any way unstable, build reinforcements into the concrete backing wall. Build the last few layers of brick or block, setting in the plastic pegs as you go, and set the inlet and outlet pipes in place. If you have got it right, the inlet pipe will run in at an angle of about 40–50° and enter the tank about 1 ft. above the base slab, while the outlet pipe will run at a much flatter angle of about 25–30° and enter the tank at a slightly lower level than the plastic dome-control pegs. Finally, render the inside walls of the tank with a coat of cement mortar.

INLET AND OUTLET PIPES AND TANKS

The inlet tank is at a higher level than the liquid in the digester tank, while the outlet tank is at a lower level. The idea is that, if the surface of the liquid in the digester is just level with the top of the outlet pipe, then for every measure of manure mix that you pour in, the same

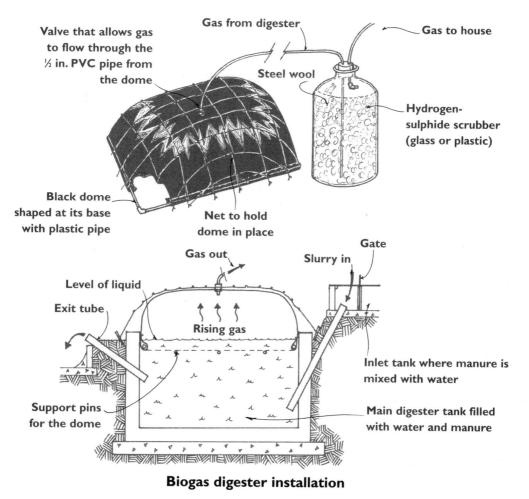

Valve that allows gas to flow through the ½ in. PVC pipe from the dome

Gas from digester

Gas to house

Steel wool

Hydrogen-sulphide scrubber (glass or plastic)

Black dome shaped at its base with plastic pipe

Net to hold dome in place

Gate

Gas out

Slurry in

Level of liquid

Exit tube

Rising gas

Inlet tank where manure is mixed with water

Support pins for the dome

Main digester tank filled with water and manure

Biogas digester installation

amount of slurry will dribble out. As for the design of the inlet and outlet tanks, you can use barrels with a large wooden plug stopping the top end of the inlet pipe, but it is much better to build both with brick and have the top end of the inlet pipe controlled with a sluice gate or valve made of wood or plastic, but not metal.

THE GAS CHAMBER AND CONTROL PEGS

The gas dome is no more than a large sheet of good-quality plastic, hemmed at the edges, and gathered onto a plastic pipe frame. The dome sits on the plastic pegs that stick out from the inside walls of the tank in such a way that the rim of the dome is a snug fit within the walls but is at the same time set a little below the level of the liquid within the tank. The gas that bubbles off from the liquid is then captured under the dome. There are other options for holding the gas dome down—you could use more pegs or a frame.

HINTS AND TIPS

Domestic biogas energy is viable—there are countless systems worldwide to prove it—but there is very little interest in the subject. Using manure may be funny, and the media never tire of this, but the fact is that if you have the space and the manure, and don't mind getting your hands dirty, a biogas system is a very good option.

THINKING IT THROUGH

- **YOUR LOCATION**

  This system needs space. The animals and the system need to be downwind, a good distance from the house, and well away from neighbors and public rights of way. Above all else, the area must be secure.

- **DO YOU NEED CITY/COUNTY PERMITS?**

  It depends, as ever, upon where you live. Always check with your local authority before proceeding.

- **HOW DANGEROUS IS IT?**

  Biogas and air or oxygen is an explosive mix and extremely dangerous. In its vicinity, you cannot smoke, work with steel-to-steel tools or power tools, use a cell phone or a radio, drive your car, move the manure with a steel bucket, or bang in a steel nail with a steel hammer. Anything that might make a spark is out. Methane can also knock you out by asphyxiation, so you cannot work in a pit that is half full of gas. There is no middle ground here—you must be qualified or you must employ qualified labor.

- **A SECURE AREA**

  While the digester must be open to the air—so that fumes and gas can waft away—it must also be 100 percent secure against unsuspecting visitors, finger-nosy children, wild animals, passing traffic, and anything or anyone else that might do harm or come to harm. For example, if a delivery man could wander up with a cigarette in hand and get close to the system, then it is not secure.

THINKING IT THROUGH (CONTINUED)

● **COMMERCIAL UNITS**

There are now some very good biogas systems on the market. They are expensive, but they come virtually preassembled—a little like a modern plastic septic tank—and are more or less ready to set in the ground.

● **WHY DOES MY DIGESTER STOP AND START?**

There are many possible reasons, but the most common problems are related to the mix ingredients, low temperatures, chemicals, and antibiotics in the mix, and the presence of large quantities of metals. Use pig, chicken, or cow manure (or from where there is a plentiful supply), avoid manure that has been contaminated with chemicals and/or medicines, don't put too much water in the mix, keep up the heat by insulating the dome, and avoid metal component parts.

● **HOW DO I GET THE WASTE SLURRY OUT OF THE TANK?**

If you have got it right, the act of putting a fresh batch of the mix down the inlet pipe will push the same amount of slurry out through the exit pipe. If you fail to get waste out, the outlet pipe may be at too high a level, and/or there is a blockage in the outlet pipe.

● **IS THE DIGESTED WASTE SAFE?**

Yes, the anaerobic process will kill the harmful bacteria. Spread the outlet waste on the vegetable garden and it will produce a bumper crop. As for harmful bacteria at the inlet end, you must make sure you are fully protected with throwaway clothes, protective gloves, and so on. Wash your hands thoroughly, and keep animals and children away from the inlet end of the digester.

Storing energy

INSULATION

One of our basic needs is shelter—protection against the wind, rain, cold, and sun. To keep your home comfortable—that is, warm in the winter and cool in the summer—research indicates that you must thoroughly insulate it. If you get your insulation right, the heat produced as you go about your day-to-day activities by the lights, the stove and your body itself will keep your home at a comfortable temperature.

WASTED ENERGY

With heating and cooling in our homes accounting for 75 percent of our energy costs, figures suggest that most of this energy leaks out through badly insulated walls, ceilings, and floors.

R-VALUES

The quality of insulation is measured in terms of "thermal resistance" (R). The better the insulation, the higher the R-value. Off-gridders should aim to push the R-value as high as it will go. A high R-value equals low heat loss, low energy consumption, and minimum energy costs.

MAKING COMPARISONS

It is almost impossible to compare like with like. For example, while foil-foam-foil (FFMF) is a wonderfully efficient insulation, its chemical make-up and production processes mark it down as a pollutant. While sheep's wool is a natural "green" product, it is expensive and needs to be used in very thick layers, which in turn increases transport costs, and so on. Natural sheep's wool has an R-value that goes up with thickness; for example, at 6 in. thick it has an R-value of 23. It is natural, sustainable, biodegradable, nontoxic, and above all, nonirritating to the skin, unlike fiberglass. Foil-foam-foil is made from a sandwich of aluminum foil and polyethylene foam, and at ¼ in. thick it has an R-value of 14.5. It is fine to use on the outside of a house but not inside because there is concern about potentially harmful gases.

ROOF AND WALL INSULATION

In a low-energy house, installing insulation is a great first step, but you must do it properly. It is no good putting a 3 in. layer of low-quality fiberglass in the attic and complaining that the insulation does not make any difference. You have to use vapor barrier membranes and breather membranes and then as much high-quality insulation as you can physically wrap and stuff over and under the floors, walls, and ceilings. Then you will see the difference.

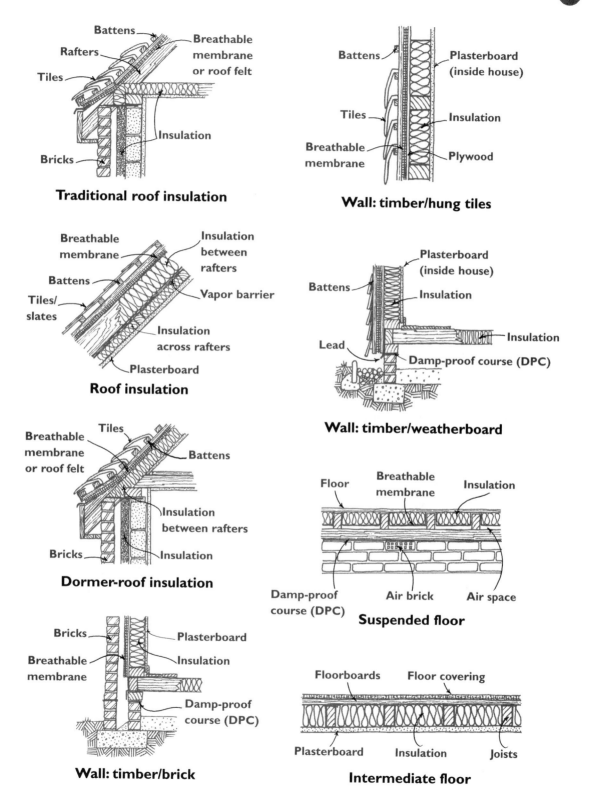

Traditional roof insulation

- Battens
- Rafters
- Tiles
- Breathable membrane or roof felt
- Insulation
- Bricks

Wall: timber/hung tiles

- Battens
- Tiles
- Breathable membrane
- Plasterboard (inside house)
- Insulation
- Plywood

Roof insulation

- Breathable membrane
- Battens
- Tiles/slates
- Insulation between rafters
- Vapor barrier
- Insulation across rafters
- Plasterboard

Wall: timber/weatherboard

- Battens
- Lead
- Plasterboard (inside house)
- Insulation
- Insulation
- Damp-proof course (DPC)

Dormer-roof insulation

- Breathable membrane or roof felt
- Tiles
- Battens
- Insulation between rafters
- Bricks
- Insulation

Suspended floor

- Floor
- Breathable membrane
- Insulation
- Damp-proof course (DPC)
- Air brick
- Air space

Wall: timber/brick

- Bricks
- Breathable membrane
- Plasterboard
- Insulation
- Damp-proof course (DPC)

Intermediate floor

- Floorboards
- Floor covering
- Plasterboard
- Insulation
- Joists

STORING WOOD

First-hand experience tells us that the whole business of wood-fired heating is a wonderfully efficient and therapeutic experience—relatively low in cost, relaxing, good for physical exercise, and altogether good fun—but only if you iron out all the wrinkles that have to do with storing wood. If you get the storage right, a warm, cozy winter will be sure to follow.

THE WOOD STORAGE

The ideal wood shed needs to be big enough for a year's supply of logs, be open-sided with a good roof over it, have an area of hard ground in front, and be roomy enough for all the sawing, chopping, and splitting associated with log fires.

SITING THE WOOD STORAGE

The wood supply needs to be nicely placed in relationship to both the source (the forest, delivery road, or gate) and the stove. On a winter's night, maybe with sleet or snow falling, slippery ice beneath your feet, and freezing winds, you need to have the wood storage close as possible to the house so that you can swiftly step out without getting variously soaked, frozen, and blasted. The shed also has to be convenient to the source.

BACK DOOR STORAGE

If there is no other choice than to have the main log storage remote from the house, then a minished outside the back door is a good compromise. All you really need is a small open-fronted shed with a projecting roof. Once a week, you move enough wood from the main supply to the back-door storage so that on the worst of the wet and windy winter nights you can quickly rush out and bring in the logs without getting wet, and without causing too much discomfort.

WOOD DELIVERY

It is vital that the delivery can be made as close to the wood storage as possible. There is nothing worse than having to move a huge load of wood from your front entrance to a shed somewhere round the back, a task that will probably have to be done in stages. Make sure that the delivery truck will have clear access, maybe via a driveway, to your storage area.

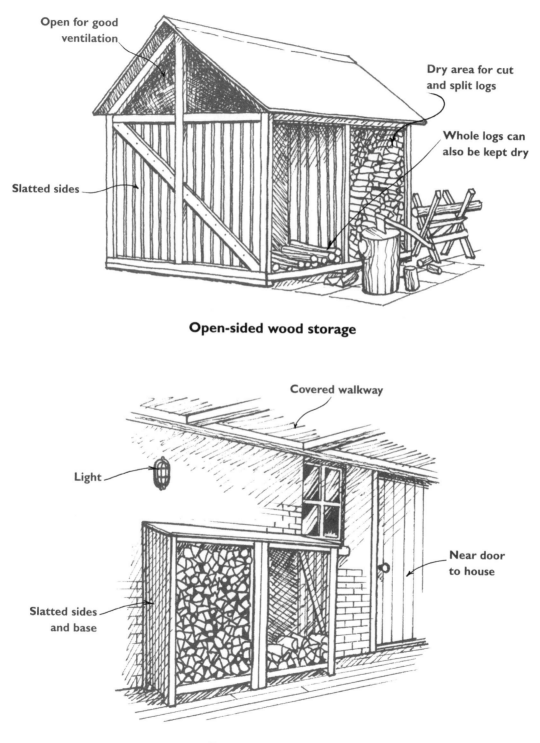

Open for good ventilation

Dry area for cut and split logs

Whole logs can also be kept dry

Slatted sides

Open-sided wood storage

Covered walkway

Light

Near door to house

Slatted sides and base

Back-door storage

STORING HEAT

If you have a solarium, you will know how quickly it heats up—one moment the sky is slightly overcast and you are sitting and enjoying a cup of tea, and the next the sun comes out and the whole place is unbearably hot. After about half an hour or so with the sun shining, most of us either throw open the doors to the yard or, if it is winter, open the doors to the house and let the heat in. Using rock or water storage takes this one step further by storing the solar heat either in an underground rock-storage bed or container or in some sort of water-storage pipe or underground tank. Either way, the sun's heat is stored during the day and given off during the night.

PASSIVE HEAT STORAGE

Passive heat storage is wonderfully simple and direct. The structure of the house (walls, floors, masonry columns, and so on) warms up during the day; at the end of the day, when the sun has gone down, the heat that is stored in the structure is, by means of windows and shutters, directed either in or out of the house. If you get it right, you can use a passive system either to heat cold areas or to draw in fresh air to cool hot areas (see pages 38–41).

ROCK STORAGE

Rock storage goes one step beyond passive solar heating; during the heat of the day the hot air in the building is directed down by means of a fan to an underground pit full of pebbles, rocks, scrap iron, crushed glass, broken brick, or other heat-retaining material. The hotter the sun, the more heat is stored. In the evening, you either reverse the fan and send warm air to various parts of the house or, if you prefer a more passive system, you open vents in the floor and let hot air rise up from the storage bed.

WATER STORAGE

A passive water-storage system—in the form of pipes, bottles, or containers of water—uses water rather than masonry to store solar heat, but is otherwise the same. Water is more efficient at storing heat than masonry, but it is heavy and it goes bad. You can mix chemicals with the water to combat the growth of algae, but there are worries about adverse effects from the chemicals. Do plenty of research before making any decisions.

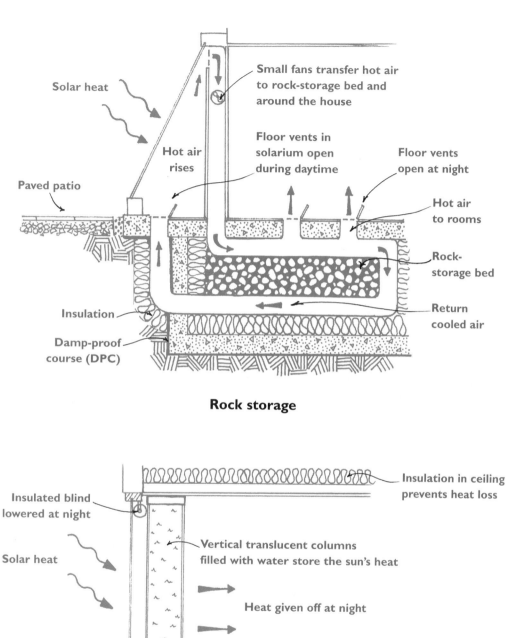

Rock storage

Water storage

STORING ELECTRICITY

In the context of off-grid energy, where you really are living without backup from the power companies, batteries are a vital factor. You need them for wind systems, photovoltaic-cell systems, for some water-turbine systems, and so on. If you are running a system where you are creating intermittent energy—the sun only shines during the day, or the wind only blows every other day, or the water only flows when you open a sluice—and you need to store it for later use, a bank of batteries is the answer.

WHAT IS A BATTERY?

In much the same way as your tank in the attic is a device used for storing water for future use, a battery, or more precisely a bank of batteries, is a device for storing electricity for future use. For off-grid energy, most batteries are of the lead-acid, deep-cycle type. The battery takes in electrical energy, chemicals within the battery change, and electrical energy is stored. When electrical energy is needed, the battery goes into reverse and gives out energy.

BEST BATTERIES FOR THE TASK

There are basically two types of batteries: deep-cycle and shallow-cycle. A cycle is one complete discharge and charge. A shallow-cycle battery is designed to give and take about 20 percent of its power in a short time, whereas a deep-cycle battery is designed to give about 80 percent of its power over a long time. You could liken a shallow cycle to one small cup of energy that is drunk quickly, and a deep cycle to a large cup of energy that is drunk slowly. With the life of the battery relating to the depth of the cycle and the number of cycles taken, it follows that a deep-cycle battery is the better option for off-grid systems where there is likely to be a long discharge and an equally long charge.

DEEP-CYCLE BATTERIES

These batteries are designed to be repeatedly "drunk dry," or discharged by 80 percent, and they tend to be large with thick, solid lead plates. They are completely unlike car batteries which are small, with sponge-like lead plates and are designed to be "sipped" many thousands of times. Certainly, you could use car batteries in an off-grid bank and deep-cycle batteries in a car—and they would work—but they would not perform efficiently and would soon fail.

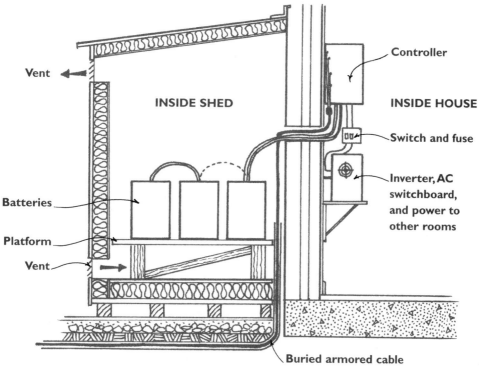

Controller

INSIDE SHED

INSIDE HOUSE

Vent

Switch and fuse

Batteries

Inverter, AC
switchboard,
and power to
other rooms

Platform

Vent

Buried armored cable
from turbine or photovoltaic array

Battery shed

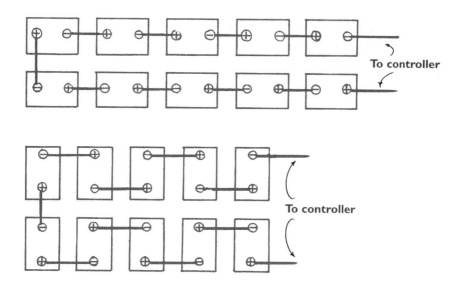

To controller

To controller

Battery-layout options for 10 batteries

BATTERY LIFETIME

How long a battery lasts depends upon the speed and depth of discharging and recharging, the way in which it is physically handled, and the temperature at which it is stored. As a very general guide, you can expect a well-treated, high-quality deep-cycle battery to last between 5 and 15 years.

THE BATTERY STORAGE

The battery shed or storage needs to be as close as possible to the house, but not so close that its proximity is in any way dangerous. Because batteries give off a potentially explosive mix of hydrogen and oxygen, are vulnerable to both high and low temperatures, and have a high scrap value, the ideal is to have them in a well-insulated, brick-built shed, set up off the floor, with generous all-around access, good ventilation at the floor and ceiling levels, and secure doors. Be warned that a bank of batteries is in every way at least as dangerous as AC household electricity, so the storage shed is no place for children, pets, open flames, gas, metal tools left lying around, or cups of water—it must be tidy and secure.

THE CONTROLLER

The function of a controller, sometimes called a regulator, is to allow the bank of batteries to become fully charged or to prevent them from becoming overcharged. In other words, if you, for example, connected a wind turbine directly to a bank of batteries—skipping the controller—you would be putting the batteries at risk and in so doing shorten their life.

THE INVERTER

The function of an inverter is to convert battery power—low-voltage DC power—into high-voltage AC or household electricity. Although there are two types of inverters in common use—low-cost modified sine wave and higher-cost sine wave—most systems recommend the sine-wave option because this produces power that best suits most appliances.

PROTECTING UNDERGROUND POWER CABLES

Underground power cables must be protected against accidental damage, such as being pierced by a pitchfork, spike, or other implement. There are two simple options: (1) you can use armored cable, a protected cable specifically designed for the job; (2) you can used ordinary cable and pass it through a protective tube that is in turn covered with a protective material such as roof tiles or bricks.

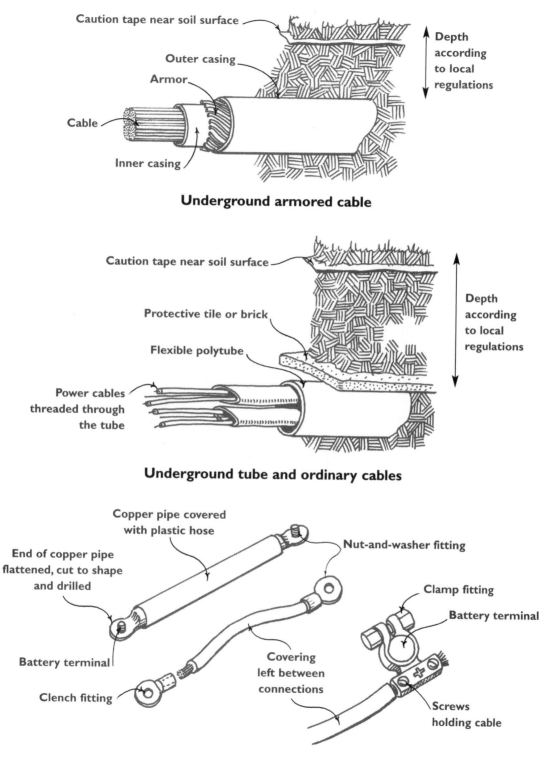

Caution tape near soil surface

Outer casing

Armor

Cable

Inner casing

Depth according to local regulations

Underground armored cable

Caution tape near soil surface

Protective tile or brick

Flexible polytube

Power cables threaded through the tube

Depth according to local regulations

Underground tube and ordinary cables

Copper pipe covered with plastic hose

End of copper pipe flattened, cut to shape and drilled

Nut-and-washer fitting

Clamp fitting

Battery terminal

Battery terminal

Clench fitting

Covering left between connections

Screws holding cable

Alternative battery connections

STORING WATER

With off-grid water, the five most common scenarios are: (1) you are living in an isolated position where a watermain is unavailable; (2) you are living in a dry area where water is precious; (3) you are plugged into a watermain supply and you are seeking to pull the plug; (4) you want to use "pure" water for personal hygiene; and (5) you are looking both to cut water costs and to become a little more "green." No matter what your motivation is, collecting and storing water has to be a good idea.

All roofs have potential for collection

Water from greenhouse roof collects in rain barrel

Linked barrels with an overflow running to pond

Trough

Patio drain to pond

Deep pond stores overflow and patio water

Water collection and storage

WATER DOWN THE PAN

Most of us use, on average, over 30 buckets of treated drinking water a day, and about one-third of this goes straight down the toilet. Therefore, the best first move you can make is to cut back on flushing the toilet in some way.

GRAY WATER

Gray water is all the used water from the shower, bath, kitchen, and laundry—but not what is put down the toilet. This equates to about 20 or so buckets a day for each person. Using this gray water to irrigate the garden, for example, would make for a substantial savings. Make sure there is no detergent in the gray water, and only use it to water plants' root systems.

STORING RAINWATER

It makes obvious sense to collect and save your own rainwater either in a simple rain barrel or in a more sophisticated system with collecting gullies, filter beds, and storage tanks (see below). The rain falls onto the roof, streams into the gullies, goes into troughs and barrels, and then on down over the filter beds and into underground storage tanks. Modern systems are made from polyplastics and are delivered to the site as prefabricated packages. If you are living in an isolated off-grid situation, or if you enjoy DIY and want to keep costs down, you could create your own concrete-and-brick system based on an old-fashioned design.

WASH-AND-FLUSH SYSTEMS

One very neat, low-cost option for cutting the amount of water that you flush down the toilet is to install a wash-and-flush system. This is a little bathroom-sink arrangement that fits on top of a low-level toilet tank. When it is filled with water, you go to the toilet, flush it, and then wash your hands in the sink. The hand-washing water then drains into the toilet tank. The next person uses the toilet, flushes with your saved hand-washing water, and washes his or her hands, and so on. It is not a huge savings, but it all adds up, and it encourages everyone to wash their hands.

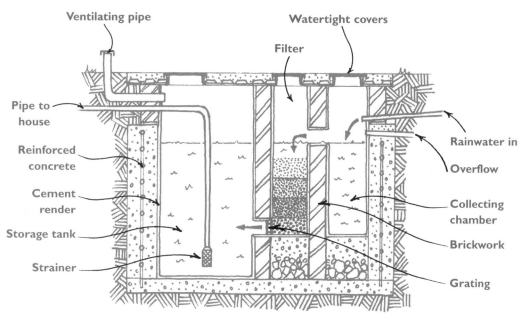

**Domestic rainwater collection-and-filtering system
(devised by Godwin and Downing)**

GLOSSARY

Alternating current (AC) Standard type of electricity used to power homes.

Alternative-energy home Home that uses non-fossil-fuel energy.

Autonomous Having the ability to function independently of other components or systems.

Biomass Plant material or vegetation used as a source of fuel or energy.

Charge controller Small electromechanical device that controls the electricity as it comes from a wind turbine, photovoltaic panel, hydroturbine or other generator. The controller ensures that the batteries are always presented with the correct voltage and current.

Damp-proof course (DPC) Layer of an impermeable material laid near the ground in the foundation walls of a building to prevent moisture from rising into the building.

Direct current (DC) Steady electric current that flows in one direction only. The current from batteries is DC.

Eco self-sufficiency Term used to describe a home and way of life that is in tune with nature.

Eco-friendly or **environmentally friendly** Not harmful or threatening to the environment.

Ecologically balanced Describes a way of life that is in harmony with the environment.

Ecology The study of the relationship between living organisms—humans, animals, and plants—and their environment—the earth.

Environmentally friendly See Eco-friendly.

Evacuated-tube collector Series of transparent glass tubes, each comprising an inner and outer tube, a heat-absorbing surface, a mirror heat-reflecting surface, and a copper heating pipe.

Fossil fuels Hydrocarbon deposits, such as oil, natural gas, and coal, that were formed underground over millions of years from the remains of dead plants and animals.

Green Popular term used to describe people, systems, groups, and ideas that are considered to be eco-friendly.

Grid Although this word was originally used to describe the power utilities—gas, electricity, and water—simply because these companies actually used grids of cables and pipes to feed our homes, it is now also used in a general way to describe any energy source that is easily distributed.

Grid-linked, **grid tie-in**, or **on-grid** Describes homes that are dependent on the power utilities. Most people in developed countries are grid-linked.

Grid tie-in See Grid-linked.

Inverter Electromechanical device that changes the DC electricity as it comes from the battery bank into AC electricity. The size, shape, and power of your inverter will relate to the size of your battery bank.

Off-grid Describes houses in the developed world that, by necessity or choice, source their own water and energy rather than tapping into the normal "on-grid" supplies of gas, electricity, and water. A house in town is usually grid-linked, while a house in a rural area, in the mountains, or deep in a forest may very well be off-grid. The term is also now used to describe an independent way of life.

On-grid See Grid-linked.

Passive solar heating System that stores the heat from the sun for later use, such as a rock-storage bed or a water-storage system.

Photovoltaic (PV) cell Panel that directly converts the light from the sun into DC electricity. Each of the cells contains a back contact, two silicone layers, an antireflective coating, and a contact grid. A batch of PV cells is capable, via an inverter, of producing a very useful amount of AC power.

Renewable or sustainable energy All non-fossil-fuel energy—wind, solar, geothermal, or biomass—that is created from a renewable source.

Self-sufficiency Self-contained system of living that involves producing food, creating energy, and recycling waste without recourse to outside agencies. A truly self-sufficient setup would produce all its own food—plants and animals—trade any excess produce, create its own energy, and manage its own waste.

Solar heater System of heating water using the sun's rays. The sun shines on a heat-absorbent surface, water within the system heats up and starts to move, the hot water is stored in a tank, the water within the tank is used or cools down and begins to circulate, and so on.

Trombe wall Named after its inventor Felix Trombe, this is a passive solar-heating and ventilation system that is made up of a masonry wall covered in glass. The sun shines through the glass, the wall absorbs heat, and the space between the glass and the wall becomes a thermal chimney. The resultant hot air within the thermal chimney is directed and channeled by natural convection either into the building so that it heats the interior, or out of the building so that it cools the interior. The joy of the Trombe system is that it can be operated without the need for complex electromechanical systems—no fans, motors, or blowers.

Turbine A motor or engine in which a wheel with vanes is made to revolve by the force of water, steam, or air. Turbines are often used to turn generators that produce electric power.

Unplugged Not connected to the utilities.

RESOURCES

ALTERNATIVE ENERGY

Alternative Energy News
www.alternative-energy-news.com

Lotus Energy (Pico-hydro)
1249 Thirbam Sadak—3, PO Box 9219,
Bhatbhateni Dhuge Dhara
Kathmandu, Nepal
Tel: 977-1-4418-203
www.lotusenergy.com

ALTERNATIVE TECHNOLOGY

Center for Alternative Technology (CAT)
Machynlleth, Powys
SY20 9AZ, U.K.
Tel: 44-1654-705953
Fax: 44-1654-702782
Email: info@cat.org.uk
www.cat.org.uk

National Renewable Energy Laboratory
1617 Cole Boulevard
Golden, CO 80401
Tel: 303-275-3000
www.nrel.gov

COMPOSTING TOILETS

Clivus Multrum
www.clivus.com

GRANTS AND FUNDING

U.S. Environmental Protection Agency
www.epa.gov

GRAY WATER TREATMENT

Oasis Design
www.graywater.net

INSULATION

Insulation 4 Less
4833 Front Street, Suite B138
Castle Rock, CO 80104-7901
www.insulation4less.com

Good Shepherd Wool Insulation
R.R. #3
Rocky Mountain Horse
Alberta T4T2A3, Canada
Tel: 403-845-6705
www.goodshepherdwool.com

MICROSOLAR THERMOSYPHON SOLAR WATER HEATERS

Solar Research Design Sdn. Bhd.
2, Jalan SS14/7F, 47500 Subang Jaya
Selangor, Malaysia
Email: microsolar@hotmail.com
www.microsolarsystem.com

SOLAR ENERGY CREDITS AND INCENTIVES

Database of State Incentives for Renewables
and Efficiency
www.dsireusa.org

Internal Revenue Service
www.irs.gov
Search for tax form 5695

U.S. Department of Energy
www.energy.gov

WASTE RECYCLING

U.S. Department of Energy
www.energy.gov

U.S. Environmental Protection Agency
www.epa.gov

WOODCHIPPERS
AND OTHER MACHINERY
Weifang Shuntiandi Import and Export
Corporation
23 Xinhua Road, Kuiwen District
Weifang, China
Tel: 86-536-8926761

WIND TURBINES
AND SOLAR POWER
American Wind Energy Association
1501 M Street NW
Suite 1000
Washington, DC 20005
www.awea.org

Aerostar Wind Turbines
P.O. Box 52
Westport Point, MA 02791
www.aerostarwind.com

Bergey Windpower
2200 Industrial Boulevard
Norman, OK 73069
www.bergey.com

Dulas Ltd
Unit 1, Dyfi Eco Park, Machylleth
Powys SY20 8AX, U.K.
Tel.: 44-1654-705000
Fax: 44-1654-703000
Email: info@dulas.org.uk
www.dulas.org.uk
Specialists in photovoltaic (PV) sources.

Jacobs Wind Turbines
Wind Turbine Industries Corp.
16801 Industrial Circle SE
Prior Lake, MN 55372
Tel: 952-447-6064
www.windturbine.net

Swift Turbines Ltd.
SAC Bush Estate
Edinburgh EH26 0PH, U.K.
Tel: 44-131-535-3301
www.renewabledevices.com

Yangzhou Shenzhou Wind Driven
Generator Co Ltd.
Xinhe Industrial Park, Xiannv Town
(Shianggou Town), Jiangdu City, Jiangsu
Province, China
Tel: 86-514-6290068
Email: yll@china-swtgs.com
www.f-n.cn

WOOD-BURNING STOVES
StovesOnLine Ltd.
Capton, Dartmouth TQ6 0JE, U.K.
Tel: 44-845-226-5754
www.stovesonline.co.uk

Woodstock Soapstone Company
66 Airpark Road
West Lebannon, NH 03784
www.woodstove.com

INDEX

ABOUT THE AUTHORS

Alan and Gill Bridgewater have gained an international reputation as producers of highly successful gardening and DIY books on a range of subjects, including garden design, ponds and patios, stone and brickwork, decks and decking, and household woodworking. They recenty moved to a small farm, so they are now writing about self-sufficiency from firsthand experience.